TRUTH, BEAUTY, AND THE LIMITS OF KNOWLEDGE

A Path from Science to Religion

To Sister Beth,
With my best wishes.
Alex Zecevic

Aleksandar I. Zecevic

University Readers™
San Diego, CA

Bassim Hamadeh, CEO and Publisher
Christopher Foster, General Vice President
Michael Simpson, Vice President of Acquisitions
Jessica Knott, Managing Editor
Stephen Milano, Creative Director
Kevin Fahey, Cognella Marketing Program Manager
Melissa Accornero, Acquisitions Editor

First published in the United States of America in 2012 by University Readers, Inc.

Trademark Notice: Product or corporate names may be trademarks or registered trademarks, and are used only for identification and explanation without intent to infringe.

File licensed by www.depositphotos.com

16 15 14 13 12   1 2 3 4 5

Printed in the United States of America

ISBN: 978-1-60927-492-4

University Readers™
800.200.3908 | www.universityreaders.com

Contents

Preface   v

I   The Known, the Unknown and the Unknowable   1

1   Faith, Reason and Analogical Thinking   3
  1.1   Analogies, Probability and Plausibility   6
  1.2   Reason and Faith   9
  1.3   Notes 13

2   Chaos, Complexity and Self-Organization   15
  2.1   What Is Chaos? 15
  2.2   Strange Attractors and Fractals 23
  2.3   Complexity and Self-Organization 26
  2.4   Notes 30

3   Metamathematics   33
  3.1   What Is a Formal System? 34
  3.2   Gödel's Theorem and Its Consequences 38
  3.3   Notes 47

4   Quantum Mechanics   51
  4.1   The State of a Quantum Particle 51
  4.2   The Uncertainty Principle 54
  4.3   The Dual Nature of Matter 57
  4.4   The EPR Paradox 60
  4.5   Notes 64

5   The Theory of Relativity   65
  5.1   Special Relativity 65
  5.2   General Relativity 69
  5.3   Black Holes 74
  5.4   Notes 80

6 String Theory   **83**
  6.1 Some Basic Properties of Strings 84
  6.2 The Explanatory Power of String Theory 87
  6.3 Notes 92

II The True, the Good and the Beautiful   93

7 Aesthetics, Science and Theology   **95**
  7.1 The Origins of the Aesthetic Drive 95
  7.2 Beauty and Truth 101
  7.3 Beauty in Mathematics and Science 106
  7.4 Notes 122

8 Ethics, Science and Theology   **125**
  8.1 Free Will and Determinism 126
  8.2 Religious Ethics and Moral Relativism 130
  8.3 The Role of Values in Science 134
  8.4 Notes 141

III Facts, Theory and Mystery   143

9 Describing the Indescribable   **145**
  9.1 Unknowable Truths 146
  9.2 The Existence of God 154
  9.3 The Attributes of God 160
  9.4 Goodness, Omnipotence and Omniscience 165
  9.5 Notes 180

10 Four Difficult Questions   **185**
  10.1 Are Miracles Possible? 186
  10.2 Can We Investigate the Unknowable? 194
  10.3 Can Religion Be Reconciled
      with Evolution? 205
  10.4 Is There a Single 'True' Religion? 216
  10.5 Notes 223

11 Epilogue   **229**

Preface

> "A little philosophy inclineth a man's mind to atheism; but depth in philosophy bringeth mens' minds about to religion." *Francis Bacon*

The book that you are about to read is undoubtedly an unusual one. When it comes to writing about science and religion, the conventional wisdom has been to minimize the presentation of scientific topics as much as possible, and avoid technical descriptions like the plague. What I propose to do is rather different. The first part of the book, for example, is devoted almost exclusively to an overview of recent results in system theory, mathematics and physics, many of which will prove to be highly counterintuitive. Science and math play a prominent role even in the sections devoted to theology, aesthetics and ethics (Chapters 7 – 11), although here they are treated mainly as a source of analogies. Having said that, however, I should add that the "uninitiated" reader has nothing to fear. Although the book does contain some diagrams and formulae, I made a concerted effort to present the relevant concepts in a way that is accessible to a wide audience.

Whether or not I have succeeded in this endeavor is, of course, for the readers to decide. But even if I have, it still remains unclear why someone whose primary interest lies in religion or the humanities should bother to learn anything about superstrings, black holes or chaos (and other similarly esoteric scientific concepts). One of my principal objectives in this Preface will be to explain why this is important, and why such individuals should read the first six chapters of the book carefully. Having given this question a great deal of thought, I can offer at least three good reasons.

1. Science and Theology as Potential Allies Contemporary skeptics have often argued that faith is incompatible with scientific knowledge, and have substantiated their claims by pointing to numerous inconsistencies and fallacies that are associated with fundamentalist views. From their perspective, religion seems to be no more than a delusion, with potentially dangerous consequences. What such critics fail to realize, however, is that these conclusions apply only to crude and simplistic interpretations of religion. Some of the claims put forth by modern theologians are, in fact, far more sophisticated, and happen to be entirely compatible with established scientific knowledge.

I would actually go a step further, and argue that building a constructive relationship between science and religion entails much more than simply avoiding contradictions. Reconciling religious teachings with existing scientific theories is clearly an important prerequisite for a meaningful dialogue, but I feel that theologians should take on an even more active role in this process. In particular, I believe that they ought to openly embrace new scientific discoveries, and learn as much about them as possible. Indeed, if we wish to build bridges across the deep gulf that separates these two domains of human inquiry, we must firmly secure the foundations on _both_ ends of the structure, and gain some familiarity with the language of the other discipline.

To get a better sense for how theology might benefit from such an approach, imagine for a moment that human knowledge is a like a balloon, which is embedded in a "sea" of unknown truths. As the balloon expands, so will its contact surface with the surrounding "mystery". When seen from a theological perspective, this metaphor suggests that the acquisition of _any_ new knowledge must be viewed as a positive development, since it brings us a step closer to the "Ultimate Mystery". It goes without saying, of course, that the true essence of this Mystery lies beyond our reach, but there is an intrinsic benefit in formulating successively better approximations of its character. Science can be a powerful ally to theologians in this enterprise, but only if they first gain a basic understanding of its most important results. In that respect, reading Chapters 2 – 6 of this book can prove to be very helpful.

2. The "Shock and Awe" Effect of Science When I teach my class on science and religion, I usually begin by telling students (who are mostly future scientists and engineers) that my primary objective is to undermine their habitual perception of reality. I refer to this approach as the "Shock and Awe" strategy. Readers who are familiar with this phrase probably know it as a military doctrine that advocates the use of overwhelming force to demoralize the enemy. My interpretation of "Shock and Awe", however, has nothing to do with warfare. I use this term in a way that is reminiscent of Niels Bohr's famous insight that "anyone who has not been shocked by quantum physics has not understood it".

Quantum mechanics is by no means the only discipline that leaves us "shocked and awed". Einstein's realization that spacetime is "curved" and "interacts" with matter has a similar effect, as do theories which suggest that all forms of matter and energy are actually composed of tiny nine – dimensional "strings". The mere possibility that nature could be organized in such an unusual manner inspires a sense of genuine wonder. This is precisely the kind of experience that theologians see as conducive to religious belief.

It is interesting to note that although some of these theories have been around for nearly a century, most people still tend to think of nature in terms of the classical Newtonian paradigm, which maintains that physical processes are generally deterministic and predictable (i.e., "well behaved"). What we have learned in recent decades suggests, however, that this is just the tip of the iceberg, and that physical reality is actually complicated, dynamic and messy, but often in a way that is strikingly beautiful. To this I might add the observation that although new facts about nature are being discovered at an ever increasing pace, its fundamental structure remains elusive, and our language for describing it appears to be hopelessly inadequate. The parallel with theology should be obvious, but in order to fully appreciate it we must first learn something about these counterintuitive scientific theories. That is what the first six chapters of this book are designed to accomplish.

3. Building a "Modern" Theological Vocabulary   In order to be intelligible (and relevant) to a contemporary audience, theology must constantly update its metaphors, and keep up with the "spirit of the times", so to speak. If this is not done, theological terminology is likely to become outdated, and may lose much of its original power and meaning. When I first began to think about this question, I was reminded of a remark that my younger son made when he was in Kindergarten. As I was waking him up one morning, he voiced his displeasure by saying: "Give me a few minutes, dad - I am booting up." If theologians wish to capture the imagination of his generation, I think they would do well to include some analogies from science and technology in their writings. That is precisely what I have tried to do throughout this book.

For this to be possible, however, it is necessary to have at least a rudimentary understanding of some of the more striking achievements of modern science. Those who make an effort along these lines will discover that some of these results are so unusual that they bring into question our most cherished beliefs about the nature of reality. This is a where science and religion appear to converge, since both claim (each in their own way) that reality is far more complex and mysterious than it seems.

How Might a Scientist Benefit from this Book?

Having spent a couple of pages discussing how a book like this might benefit readers who are interested in theology and the humanities, I must now say a few words about what it can offer to those who already possess a solid mathematical background. Chapters 7 – 11, which are devoted to aesthetics, ethics and theology, will allow technically savvy individuals to acquaint themselves with certain insights that are seldom discussed within the scientific community. Among these insights, I would like to single out three that are particularly instrumental in establishing a constructive dialogue between science and religion.

1. The Role of Beauty in Science   Although many physicists, mathematicians and engineers recognize the value of beauty in their work, relatively few among them actually ponder why aesthetic criteria provide such an effective framework for new discoveries. The fact of the matter is that we have no good reason to expect that our sense of beauty (which seems to be a subjective category of the human mind) should be an appropriate guide to "objective" truth. There is an element of true mystery to this phenomenon, which naturally invites a conversation with theology.

2. The "Responsiveness" of Nature   An additional impetus for a meaningful dialogue between science and religion is the assertion that physical reality appears to have a "responsive" side, which can be directly influenced by human decisions. Quantum mechanics has established, for example, that if we choose to perform one type of experiment, an electron will exhibit the characteristics of a material particle. If, on the other hand, we elect to change the experimental setup, we can ensure that it will behave like a wave (which is a very different type of "physical reality"). It seems, in other words, that we can determine which "face" nature will show us simply by choosing how we will observe it. If that is indeed the case, then there is clearly room for discussions that go well beyond mere theory and experiment, and delve into the very nature of reality. For those scientists who are open to religion, this realization is undoubtedly a welcome development (as are many other counterintuitive claims of modern physics). I strongly believe that the theological aspects of this book will provide such individuals with an appropriate framework for exploring the underlying metaphysical and spiritual implications.

3. Scientific Metaphors as a Way of Interpreting Theological Claims

Even if we agree that theological insights can benefit open-minded scientists, there is still a nontrivial language barrier that needs to be overcome. This barrier is a result of the fact that science and theology have been evolving independently for a very long time, and have developed very different vocabularies. My own way of dealing with this problem has been to "translate" difficult theological concepts and interpretations into the more familiar language of mathematics and science. Over the past 15 years, such analogies proved to be extremely helpful, to the point that they have now become one of my most potent tools for bridging the two disciplines. I suspect that other scientifically minded individuals might find this methodology equally fruitful as they examine their own attitudes toward religion.

The Historical Context for This Book

In order to place this book into the proper historical context, it is important to keep in mind that I am by no means the first to suggest that science and religion are connected on a deeper level. The notion that these two disciplines are essentially inseparable has a long history, which dates back (at least) to ancient Greece and the school of Pythagoras. According to the Pythagoreans, mathematics was not just a way to understand nature, but also a key to recognizing certain profound spiritual truths. Numbers and geometric forms were believed to have a symbolic meaning, which points beyond the physical world to a transcendent and eternal reality.

Such views had a profound impact on western thought, and even inspired a "sacred geometry" that was subsequently used in the construction of many religious structures. The proportions of the Parthenon, for example, were chosen to emulate the so called Golden Ratio ($\phi : 1 \approx 1.6180339887\ldots$), which was thought to be a number with a special mystical significance. Over the centuries, this ratio attracted the interest of many great thinkers, artists and scientists, including the likes of Leonardo da Vinci, Fibonacci, Roger Penrose, Le Corbusier and Salvador Dali (to name just a few). In his book *The Golden Ratio*, Mario Livio describes our enduring fascination with ϕ in the following way:

> "The fascination with the Golden Ratio is not confined just to mathematicians. Biologists, artists, musicians, historians, architects, psychologists, and even mystics have pondered and debated the basis of its ubiquity and appeal. In fact, it is probably fair to say that the Golden Ratio has inspired thinkers of all disciplines like no other number in the history of mathematics."

Although science and religion continued to be intertwined throughout the Middle Ages, during the Renaissance these two disciplines began to drift apart. Perhaps the last great theologian who tried to fully integrate scientific and religious knowledge into a seamless whole was cardinal Nicholas of Cusa (1401 – 1464). In the course of his illustrious career, he wrote treatises on theology, philosophy, geometry, logic and astronomy, and was considered to be one of the most original thinkers of his time. He is perhaps best known for his studies of mathematical infinity, which foreshadowed the discovery of calculus, and even anticipated Cantor's ground-breaking work on set theory in the 19th century. Nicholas of Cusa was also one of the first to suggest that the earth is not the center of the universe and that heavenly bodies do not move in perfect circles. It is believed that this conjecture laid the foundation for the subsequent discoveries of Giordano Bruno (1548 – 1600) and Johannes Kepler (1571 – 1630), both of whom were familiar with his work.

What I find particularly interesting about Nicholas of Cusa was the way in which he used science as an analogical framework for interpreting religious teachings (an approach that I favor as well). Much of his theological work was based on the premise that human knowledge has fundamental limits, which neither science nor mathematics can overcome. We now know that such limits do, in fact, exist, thanks to recent discoveries in quantum mechanics, metamathematics and chaos theory. For Nicholas of Cusa, however, this was primarily a theological conjecture, which he justified analogically by comparing truth to the notion of mathematical infinity. He argued that both can be approximated, but never attained by finite means.

Sadly, such an "integrative" approach to knowledge has now been largely abandoned, and science and theology have evolved into very different (and seemingly unrelated) disciplines. This separation seems to be a perfect reflection of contemporary society. As philosopher Will Durant once remarked, we live in a specialized world where it is good to "know more and more about less and less." In such an environment, any form of integration across disciplines carries little weight outside academic circles, and even there it is "unsafe" to venture too far beyond the boundaries of one's formal area of expertise. Whenever that is done, errors and misinterpretations become increasingly more likely. I do believe, however, that the time may be right to pick up where Nicholas of Cusa left off some five centuries ago. According to the British embryologist (and philosopher of science) C. H. Waddington:

"The acute problems of the world can be solved only by whole men, not by people who refuse to be, publicly, anything more than a technologist, or a pure scientist, or an artist. In the world of today, you have to be everything or you are going to be nothing."

If Waddington's assessment is correct, it is reasonable to expect that the complexities of the modern world will force us to think once again in broadly interdisciplinary terms, and attempt to connect widely disparate fields of human knowledge (including, of course, science and theology). Perhaps this has always been a natural human impulse, which was held in check by our professional vanity and the fear of error. If that is the case, however, we must concede that these limitations are largely self-imposed, and that it is clearly within our power to transcend them. This may not be an easy thing to do, and may result in a few mistakes along the way. But I have no problem with that. Ultimately, we can always respond to our critics with Woody Allen's classic line from Annie Hall: "You mean my whole fallacy is wrong?"

How to Read this Book

As noted at the beginning of this Preface, Chapters 2 – 6 contain an overview of some striking and highly counterintuitive results of modern science. Those who feel that their technical background is "weak" might consider skimming through some of the more challenging material, and can focus their attention on the theological and philosophical questions that are addressed in Chapters 7 – 11. In examining these questions, I have used scientific results mainly as a source of analogies, which require only a rudimentary understanding of the underlying physical theories. This part of the book also includes a discussion of aesthetics and ethics (in Chapters 7 and 8), since these two disciplines are of significant interest to both science and theology. In these chapters I have tried to demonstrate that science and religion share certain fundamental values, which can sometimes counterbalance the profound differences that separate them.

It goes without saying, of course, that scientific analogies must be used with great care in theology, since religious truths cannot be reduced to formal laws or theorems. The following conversation from Huxley's *Point Counter Point* exposes the absurdity of such attempts.

"This is me. Edward, I've just discovered a most extraordinarily mathematical proof of the existence of God, or rather of ..."

"But this isn't Lord Edward," shouted Illidge. "Wait. I'll ask him to come." He turned back to the old man. "It's Lord Gattenden," he said. "He's just discovered a new proof of the existence of God." He did not smile, his tone was grave. Gravity in the circumstances was the wildest derision. The statement made fun of itself... "A mathematical proof," he added more seriously than ever.

"Oh dear!" exclaimed Lord Edward, as though something deplorable had happened. Telephoning always made him nervous. He hurried to the instrument. "Ah, Edward," cried the disembodied voice of the head of the family from forty miles away at Gattenden.

"Such a really remarkable discovery. I wanted your opinion on it. About God. You know the formula: m over nought equals infinity, m being any positive number? Well, why not reduce the equation to a simpler form by multiplying both sides by nought? In which case you have m equals infinity times nought. That is to say that a positive number is the product of zero and infinity. Doesn't that demonstrate the creation of the universe by an infinite power out of nothing? Doesn't it?"

In writing this book, I have made a concerted effort *not* to sound like Lord Gattenden. It is largely for this reason that I decided to devote considerable attention to science and mathematics, and use them as an analogical framework for analyzing certain important theological claims. Most of the arguments that I have developed along these lines are confined to Christianity, for the simple reason that I am not sufficiently familiar with other traditions. In that sense, there is an autobiographical element to this book, since it represents a summary of the questions and possible answers that I have encountered in discerning my own faith (I am Serbian Orthodox). The nagging dilemma, of course, is whether I actually have anything new to say on these topics. That would be a tall order indeed, given that these issues have been explored for several millennia by minds far greater than my own. I am heartened, however, by the advice of C. S. Lewis:

"In literature and art, no man who bothers about originality will ever be original; whereas if you simply try to tell the truth (without caring how often it has been told before) you will, 9 times out of 10, become original without ever having noticed it."

I suppose that in the final analysis it must be conceded that very few among us are capable of generating completely new ideas. But what remains uniquely our own is the way in which we connect the existing ones.

Acknowledgements

My work on this book was supported by grants from the Center for Science, Technology and Society (CSTS) and the Markkula Center for Applied Ethics at Santa Clara University. I am grateful to both Centers for their generosity and encouragement, and especially to Dr. Geoffrey Bowker (the former Director of the CSTS), whose openness to new ideas and willingness to promote interdisciplinary research have played an important role in this project.

A number of my colleagues read the manuscript (or parts of it) and provided many constructive suggestions. I am particularly indebted to the following individuals, whose varied expertise reflects the broad scope of this book: Ruth Davis (Computer Engineering), Timothy Healy, Radovan Krtolica, and Allen Sweet (Electrical Engineering), Betty Young (Physics), Tracey Kahan (Psychology), Mark Graves (Religious Studies), Andre Delbecq (Management), Alejandro Garcia-Rivera (Jesuit School of Theology), Judith Dunbar (English), Lancelot Pereira, S.J. (Life Science and Biochemistry - Xavier Institute of Engineering, Mumbai) and Robert Audi (Philosophy - Notre Dame University). I also greatly appreciate the valuable feedback that I received from members of the reading groups in Theological Ethics, Mystical Theology and Interdisciplinary Aesthetics, as well as the participants in the Ignatian Faculty Forum. Needless to say, the views and opinions expressed in this book are my own, as are any errors.

Finally, I owe a special debt of gratitude to my wife Jelena, who has been my greatest supporter and toughest critic throughout this endeavor. She read countless versions of the manuscript, and has helped make it far better than it originally was.

Part I

The Known, the Unknown and the Unknowable

Chapter 1

Faith, Reason and Analogical Thinking

"Life has taught me how to think, but thinking has not taught me how to live." *Alexander Herzen* [1]

"The value of speculative answers, however judicious, is limited. They clear the way for an apprehension of truth which speculation alone is powerless to reach. Peasants and housekeepers find what philosophers seek in vain; the substance of truth is grasped not by argument, but by faith." *Austin Farrer* [2]

At some point in life most of us begin to wonder about the purpose and meaning of our existence. More often than not, these questions arise when we face situations that force us to recognize the full extent of our limitations. Such encounters with reality can be overwhelming, and can lead to a reexamination of one's attitude toward religion. Scientists are certainly not exempt from such experiences, but what makes their response rather unique is an intense need for reconciling faith with reason. This need follows from the very nature of their profession, which allows little room for beliefs that have inadequate logical support. Such beliefs are usually rejected as irrational, or are (at best) accepted with a great deal of skepticism.

It is fair to say that the standards of rationality set by science have become prevalent in contemporary society. To a large extent, this trend reflects the gradual secularization of the Christian world which began with the emergence of modern science in the early 17th century. Over the past four hundred years, this process produced a deep chasm between science and religion, which has become increasingly difficult to bridge. The challenge is particularly daunting when it comes to the deep mysteries of faith, which often defy the basic concepts and categories of analytical thinking. When applied to such cases, conventional reasoning has little to contribute, and often leads to paradoxical conclusions. This apparent inconsistency has led to the rather widespread view that religious teachings lack coherency, and are therefore unacceptable to scientifically minded individuals.

It would be wrong, of course, to assume that the position outlined above is universally accepted. Some prominent scientists (such as evolutionary biologist Steven Jay Gould, for example) have sought a more moderate stance, which sees theology as a legitimate domain of human inquiry with its own method and logic. Thinkers of this persuasion emphasize, however, that such an approach to reality is completely unrelated to science, and that the two cannot be connected in any meaningful way. This outlook is usually supported by the observation that science deals with empirical facts and mathematical models, while religion addresses a very different set of questions, which primarily relate to meaning and value. Since the two fields appear to have little in common, it follows that a well informed and thoughtful person could conceivably engage in both without encountering serious logical difficulties.

> "I do not see how science and religion could be unified, or even syn-
> thesized, under any common scheme of explanation or analysis; but
> I also do not understand why the two enterprises should experience
> conflict." *Steven J. Gould* [3]

Gould's idea of "non-overlapping magisteria" strikes a rather conciliatory note, and seems to provide a sensible compromise in the debate between science and religion. The question, however, is whether such a neutral position is sustainable in practice. Several thinkers have been quick to point out that faith constitutes a global view of reality, and that its pronouncements must necessarily apply to nature as well. They have also emphasized the fact that science is inherently incomplete, since it is incapable of answering certain fundamental questions about the universe and the laws that govern it. Astrophysicist Robert Jastrow argues that such questions represent a natural point of contact between science and theology:

> "At this moment it seems as though science will never be able to
> raise the curtain on the mystery of creation. For the scientist who
> has lived by his faith in the power of reason, the story ends like a
> bad dream. He has scaled the mountains of ignorance; he is about
> to conquer the highest peak; as he pulls himself over the final rock,
> he is greeted by a band of theologians who have been sitting there
> for centuries." [4]

While Jastrow makes a legitimate point regarding the limited scope of scientific inquiry, it is important to recognize that religious teachings have their own set of constraints. Theologians will readily concede, for example, that the language they use to describe God is hopelessly inadequate. Many of them will also agree that theological claims must meet certain standards of rationality in order to be intellectually acceptable to an educated population. These standards imply (among other things) that our beliefs about the nature of reality should not contradict established scientific knowledge.

"Theological doctrines must be consistent with scientific evidence even if they are not directly implied by current scientific theories." *Ian Barbour* [5]

It would seem, then, that a certain amount of interaction between science and religion is unavoidable, and that the two disciplines cannot be completely separated. At the very least, they can scrutinize each other's claims, and serve to moderate extreme positions on both sides. In a more optimistic scenario, they might even act as an integrating factor for society. This is presumably what Pope John Paul II had in mind when he wrote that:

"We must ask ourselves whether both science and religion will contribute to the integration of human culture or to its fragmentation. ... A simple neutrality is no longer acceptable. We are asked to become one. We are not asked to become each other." [6]

The view expressed by John Paul II stresses the need to examine whether science and religion can coexist, and perhaps achieve some measure of cooperation. From a scientific standpoint, the answer to this question depends to a large extent on whether religious teachings can be perceived as *rational*. Given that this issue is a central theme in the chapters that follow, it seems appropriate to begin by examining what the attribute 'rational' actually entails in this context. Does it mean that each theological statement must be accompanied by a rigorous logical proof before we can accept it? This would be an unreasonable requirement by any standard.

"A proof, I suppose, is something that will convince anyone who is intelligent enough to understand it. If so, very little of interest regarding major philosophical issues can be proved ... So, if we demand proofs in philosophy, we will wind up as skeptics on all or nearly all of the important issues." *C. Stephen Layman* [7]

Layman's observation underscores the fact that beliefs are acquired in many different ways, and that pure logic and formal proofs are by no means the only acceptable techniques for resolving complex questions. It is interesting to note that this claim holds true in the domain of science as well. Indeed, it is well known that certain "informal" criteria such as simplicity and beauty have always played a key role in the process of scientific discovery, as have intuition and imagination.

"No scientist thinks in formulae. ... The words of the language, as they are written or spoken, do not seem to play any role in my mechanism of thought. The physical entities which seem to serve as elements in thought are certain signs and more or less clear images which can be 'voluntarily' reproduced and combined. ... Conventional words or other signs have to be sought for laboriously only in a second stage, when the mentioned associative play is sufficiently established and can be reproduced at will." *Albert Einstein* [8]

If we acknowledge (as Einstein did) that science is much more than a collection of impersonal and purely objective statements, it would make little sense to define rationality in strictly "formal" terms. What we need instead are criteria that are sufficiently flexible to capture the human experience in all its diversity.

1.1   Analogies, Probability and Plausibility

In thinking about what constitutes a "rational" belief system, it is useful to recognize that our actions and opinions are guided by what physicist (and historian of science) Gerald Holton describes as a "robust, map-like constellation of ... beliefs about how the world as a whole operates." He refers to this overall outlook that shapes our attitudes as our *Weltbild* (which is a somewhat broader German term for "world view") [9]. Although the *Weltbild* of any given individual depends to a large extent on his social, ethnic and educational background, it is fair to say that it always contains a subset of beliefs that pertain to the natural world. It is perhaps here that we might locate an appropriate meaning for the attribute "rational," at least when it comes to scientifically minded individuals. I would suggest that for such a person, a coherent world view would be one that satisfies the following two conditions:

1) The set of "core" beliefs about the natural world must to be compatible with existing scientific knowledge.

2) The "non-scientific" core beliefs should be consistent (at least in some measure) with the scientific ones.

In applying these criteria, it is important to keep in mind that the term "consistent" must be used somewhat loosely. Indeed, I seriously doubt that the entire mindset of any individual could pass a strict test of logical soundness (which is perhaps what makes us human in the first place). With that in mind, I would argue that the rationality of our *Weltbild* can be justified by establishing appropriate "logical bridges" between the disparate clusters of views that constitute it. Formal proofs are of little use in this enterprise, and should ultimately give way to *analogies* and *metaphors*.

What is it about analogies and metaphors that makes them so suitable for this purpose? Our primary motive for focusing on these two modes of description stems from the fact that they have always been a natural tool for explaining difficult concepts, both in science and in theology. If these two disciplines are seen as manifestations of the *same* overarching reality (as Christian theology suggests), then it is perfectly reasonable to assume that analogies can also help bridge the apparent gap that separates them. From a theological perspective, what we are really proposing here amounts to adding a certain number of "scientific" metaphors to the already existing traditional ones. The potential value of such metaphors has been recognized by several contemporary thinkers:

> "Metaphors 'fund' theology, providing the language and images out of which theological concepts grow; they describe the unknown in terms of the known. ... When metaphors lose their original meaning

and fruitfulness, the theology built upon them must be reconstructed, drawing upon new metaphors appropriate for a new age... It seems reasonable that physics, as well as biology and the other sciences which infuse our culture, can be a source of religious metaphors." *Robert J. Russell* [10]

"For religion to be able to meet the needs of its day (to answer the question of meaning) it must be in accord with, and understandable in the language of, the scientific knowledge of the time." *Wallace Clift* [11]

Before proceeding down this path, it is worth considering whether analogical interpretations carry any weight in the domains of science and mathematics (beyond the indisputable educational value that they posses). It turns out that they do. An obvious example of this kind arises in the context of formal logic, where "interpretations" allow us to establish the consistency of abstract systems, and determine the truth (or falsity) of purely symbolic statements. To get a sense for how this works, consider the apparently random string of characters NNNPNQNN and ask yourself whether this can be a "true" statement in some formal system. There is obviously no way to tell, unless we attach some *meaning* to these symbols. The interpretation provided in Table 1.1 suggests one possibility, which allows us to "translate" our string into a true statement about integers (*i.e.* $3 + 1 > 2$). Based on this analogy between the symbols N, P and Q and the elements of number theory, we can conclude that NNNPNQNN can indeed be a "theorem" (the formal basis for this claim is provided in Chapter 3).

Symbol	Interpretation
P	$+$
Q	$>$
N	1
NN	2
NNN	3

Table 1.1. A possible interpretation of the symbols.

Although such a precise procedure is clearly impossible to reproduce in the case of theology, one could nevertheless argue that analogies perform a similar function. Indeed, their primary purpose is (and always has been) the "translation" of complex and counterintuitive theological claims into a more manageable framework. If this framework happens to be math or science, a careful choice of analogies could provide some new (and possibly unexpected) insights about certain religious beliefs, and could perhaps help us reconcile them with the scientific elements of our *Weltbild*.

Analogies between scientific and theological concepts can also be used to establish the *plausibility* of certain contentious religious propositions, which is a necessary step in justifying their rationality. To see why 'mere' plausibility is sufficient for this purpose, we must make a brief excursion into probability theory, and examine how cumulative evidence can affect our evaluation of competing hypotheses. We begin with the observation that the notion of probability is often related to the *relative frequency* with which certain outcomes occur. When we say, for example, that the probability of drawing a particular card from a standard deck is $P = 1/52 = 0.019$, it is implicitly assumed that the experiment will be repeated many times (which is typically the case in science). Thus, if we were to attempt 1,000 independent draws, the probability calculated above suggests that we should expect 19 of them to produce, say, the ace of spades.

It is obvious, however, that this interpretation doesn't work particularly well in the domain of religion, where non-repeatable situations are a common occurrence. In order to extend the notion of probability to phenomena that do not lend themselves to repeated experimentation, we need to think of it as a measure of how strongly we believe in a proposition (or a hypothesis). Any such evaluation must necessarily take into account available evidence, and the fact that probabilities are different *before* and *after* an event occurs. These probabilities are related by *Bayes' theorem*, which states that

$$P(H|E) = P(E|H) \cdot P(H) \tag{1.1}$$

where:

1) $P(H)$ is the probability that hypothesis H is true *before* event E occurred (the so-called *prior* probability).

2) $P(H|E)$ is the probability that H is true *after* event E occurred (also known as the *posterior* probability).

3) $P(E|H)$ is the probability that event E will take place *if* H is true.

To gain a better understanding of what this theorem really means, let us consider a simple example which examines whether it is 'reasonable' to believe in the existence of the Loch Ness monster.[12]

Example 1.1. In the case of the Loch Ness monster, there appear to be only two possible hypotheses:

H_1 : The monster exists.

H_2 : The monster does not exist.

If we are inclined to believe (as many are) that the monster story is no more than a myth, we will assign a small probability to hypothesis H_1. For illustrative purposes, suppose we have decided that $P(H_1) = 0.01$ and $P(H_2) = 0.99$, which is equivalent to assuming that the odds are initially $99 : 1$ in favor of hypothesis H_2. What kind of evidence would be needed to substantially change these odds? To see that, let us assume that a moving serpent-like shape was observed (and photographed) in the lake by a team of scientists who happened to be flying over in a helicopter. In the following, we will refer to this sighting as event E_1.

In order to apply Bayes' rule, we must first determine probabilities $P(E_1|H_1)$ and $P(E_1|H_2)$ (*i.e.* the probabilities that event E_1 would take place given each of the two hypotheses). If we assume that the monster *exists*, it is by no means surprising that event E_1 should occur, which means that we can assign a rather large value to probability $P(E_1|H_1)$. If, on the other hand, we assume that there is *no* monster, the sighting would be quite surprising, and we could associate a fairly low value with $P(E_1|H_2)$ (not too low, however, since we must allow for optical illusions, faulty equipment, *etc.*).

For the sake of argument, let us suppose that $P(E_1|H_1) = 0.9$ and $P(E_1|H_2) = 0.1$. According to Bayes' theorem, the probabilities of the two hypotheses *after* event E_1 took place will change to $P(H_1|E_1) = 0.009$ and $P(H_2|E_1) = 0.099$, respectively, and the odds become $11 : 1$ in favor of H_2. Based on these odds, it is still much more likely that the monster is a mythical creature, but the discrepancy between the two hypotheses is dramatically lower than it initially was.

Let us now consider a scenario where not one, but rather *two* independent events occur: E_1 (the same event that was described above) and E_2 (a large moving object has been detected at the bottom of the lake using a sonar). Since there is no correlation between the two events, it can be shown that the odds are now reduced to only $1.2 : 1$ in favor of H_2. In other words, after these two observations (each of them inconclusive), the two hypotheses have become almost equally likely, despite the very large initial difference.

What can we conclude from this example? The "moral" of the story is that cumulative evidence can dramatically alter the odds in favor of some hypothesis, no matter how unlikely it may have seemed at the outset. The only scenario where this conclusion *fails* to apply is when $P(H) = 0$. This would be the case, for example, if the hypothesis is self-contradicting, or is in direct conflict with established scientific facts. Under such circumstances, all evidence becomes irrelevant and can be dismissed a priori.

Arguments based on the fact that $P(H)$ can be zero have been quite popular among secular thinkers, who maintain that some religious claims are simply too outrageous, and cannot be reconciled with established scientific knowledge. It is largely for this reason that it is fundamentally important to show that certain counterintuitive theological teachings could *conceivably be true*. This is where analogies with science can be particularly useful - they can help establish the *plausibility* of such propositions, thereby ensuring that we can assign a non-zero prior probability to them. Then (and only then) can we include evidence such as historical testimonials and accounts of mystical experiences, and use it to justify the rationality of religious beliefs.

1.2 Reason and Faith

Although the probabilistic model outlined above provides a reasonably system-atic framework for analyzing religious hypotheses, it would be entirely inap-propriate to assume that theoretical conclusions of any kind (however logically sound they may be) can provide a proper foundation for religious belief.

> "Philosophical thinking can enable us to see through objections to Christian belief .... but it rarely, if ever, propels one into a condition of faith." *William P. Alston* [13]

> "You can find truth with logic only if you have already found truth
> without it." *G. K. Chesterton* [14]

I am inclined to agree with Alston that religious belief is not something that
can be based on formal proofs. Faith, like love, is first and foremost an *experience*, which is rarely (if ever) the result of careful analysis. I also consider it
quite unlikely that skeptics and atheists can be swayed by arguments demonstrating that religion is *not* inherently irrational. However, for those who believe
(or wish to believe), it is important to recognize that logic and faith cannot be
completely separated. There are at least two reasons why this is so. To begin
with, we should observe that much of what we do and desire is based on our core
belief system (*i.e.*, our *Weltbild*), which clearly includes our attitudes toward religion. In deciding between different courses of action we generally consult these
beliefs, so their rationality is clearly relevant. It is difficult to envision how a
distorted view of reality could be conducive to the fulfillment of our purposes,
however noble these may be.

> "It is not enough to feel deeply; one must also know. Deep care without concomitant skills and knowledge leads only into enthusiasms."
> *Michael J. Buckley* [15]

Reason also plays an important role in *sustaining* faith. To see this more
clearly, it is helpful to draw an analogy with the way in which we experience
love. There is no doubt that philosophical arguments cannot initiate this feeling,
but it is also true that love with no rational support is necessarily incomplete.
Indeed, if you have no idea why you love someone, your feelings are likely to be
transitory and not particularly flattering to the object of your affection. This
sentiment is nicely illustrated by one of my favorite literary rejections. In Jane
Austen's *Pride and Prejudice*, Elizabeth Bennett flatly turns down Mr. Darcy's
offer of marriage with the question:

> "I might as well inquire ... why you chose to tell me that you liked
> me against your will, against your reason, and even against your
> character?" [16]

If we accept, then, that religious belief cannot be disassociated from reason,
this sets certain standards regarding what can be believed. It becomes impossible, for example, to allow for superstitious beliefs and practices that often
accompany legitimate theological claims. And while we will argue in the following chapters that miracles may be logically possible, it would certainly not be
prudent to believe every such report. Even the way in which we envision God
can sometimes be thoroughly misleading and rationally unacceptable.

> "When the Christian tradition speaks of God, it does not mean a
> great big person out there somewhere, older, wiser, stronger, than
> you and I. That is Zeus, not God." *Michael J. Himes* [17]

The task of reconciling reason and faith is further complicated by the fact
that most religious traditions entail specific beliefs that go well beyond the mere

existence of God. Christians, for example, are supposed to believe in the doctrine of the Trinity, the immortal soul and the Resurrection. Teachings like this are difficult to justify on rational grounds, and incorporating them into a scientific world view poses formidable challenges. And yet, many scientifically minded individuals *do* believe, despite the apparent contradictions. It is therefore clearly of interest to examine whether such a position is logically defensible. A partial answer to this question may perhaps be found with the help of *decision theory*, which is an area of mathematics that deals with decision making under uncertainty. One of the most important results of this theory is that a rational course of action ought to be based not just on the *probabilities* of different propositions, but also on the potential *rewards* and *penalties* that are associated with them.

To illustrate this point more clearly, suppose that you have the opportunity to take a luxurious vacation on a remote exotic island (all expenses paid). Suppose further that the trip is scheduled at a time of year when mosquitoes are fairly active, and that the best available protection in the resort are mosquito nets on the beds. If we assume that the chances of contracting a nasty tropical disease are no more than 1%, the odds alone would suggest that there is no compelling reason not to go. In making such a decision, however, we must also compare the potential rewards of the trip with the potential dangers, and take into account the discomfort associated with malaria, dengue fever, and a host of other possible ailments. In view of these additional considerations, our decision may easily change, and we may choose to pass on the opportunity. In other words, it would be entirely rational to conclude that the possible penalties actually outweigh the seemingly low odds that one of the risks will materialize.

The line of reasoning used above can obviously be extended to religion as well, since belief in its teachings necessarily involves an irreducible element of uncertainty. Here, too, a rational choice would be one that weighs the risks and rewards, while taking into account all the relevant likelihoods. It goes without saying, of course, that this is *not* how such decisions are usually made. In these matters emotions and experience have far more influence than theoretical considerations. Nevertheless, it is fair to say that decision theory can be used to justify religious beliefs *after* they have already been acquired. From that perspective, a believer who happens to value the potential rewards of communion with God far more than the risk of spending a lifetime in the service of an illusion has made a perfectly rational choice (even if she sees the odds in favor of the "religious" hypothesis as fairly low).

In examining the rationality of faith, it is also useful to recognize that spiritual development is a continuous process, which generally evolves through a number of distinct stages. In each stage our perception of religion and its doctrines is qualitatively different, becoming progressively less literal as we move to higher levels. Although it is impossible to model this process with any precision, the following hierarchy suggested by M. Scott Peck may be a helpful tool for further discussion. [18]

1. *The Chaotic Stage.* This is a stage of underdeveloped spirituality, in which people are self-centered to the point where they are essentially incapable of caring for others. All very young children pass through such a stage,

but there is a non-negligible number of adults who remain entrenched in it. It is quite possible for such people to be professionally successful – they can even assume prominent leadership roles.

2. *The Formal Stage.* This level of spiritual development is characterized by belonging to a structured religious organization, typically a church. People in this stage tend to get attached to forms and rituals, often without examining their deeper meaning. The underlying belief system is taken for granted, and there is no tolerance for change or expressions of doubt.

3. *The Skeptical Stage.* In the skeptical stage there is a tendency to search for truth in a purely rational manner. Not surprisingly, many scientists and engineers fall into this category. People in this stage often question their fundamental beliefs, and are in many ways closer to secularism than to faith. Nevertheless, they can be viewed as more spiritually developed than those who remain in the formal stage.

4. *The Mystical Stage.* For those in Stage 3 who have devoted enough time and energy to the pursuit of truth, there usually comes a point when they begin to recognize some of the limitations of the scientific method, and the inability of reason to cope with certain fundamental questions. This is a point at which it becomes possible to return to religion, but in a very different way from the one associated with Stage 2.

"While Stage 4 men and women enter religion in order to approach mystery, people in Stage 2, to a considerable extent, enter religion in order to escape from it." *M. Scott Peck* [19]

Although the four stages are quite different, they are not easily distinguished in practice. In moments of fear, for example, we may well return to Stage 1; at other times, the comfort and security of Stage 2 is more appealing than the cold rationality of Stage 3. What matters, however, is the predominant state of consciousness, which should presumably evolve in a way that favors the mystical stage. It is this gradual transition that leads to what I see as an 'intelligent faith', a spirituality that *includes* and *transcends* knowledge.

For a scientist whose mindset revolves around reason and logic, the first step toward this "intelligent faith" usually involves the recognition that there are intrinsic limits to what is knowable. Theologians and atheists alike maintain that these constraints open the door to religious belief.

"The limiting of reason is a necessary ingredient for the concept of faith; it is what makes the concept of faith possible." *George Smith* [20]

"Imperfect knowledge is of the very essence of faith." *St. Thomas Aquinas* [21]

With that in mind, a significant portion of this book is concerned with exploring the limits of science and mathematics, together with their theological and philosophical implications. A deeper understanding of chaos theory, meta-mathematics, quantum mechanics and relativity can be very helpful in that respect, and can provide a natural framework for speaking about the rationality of faith. In the chapters that follow, we will examine these four scientific topics in greater detail, and will outline how each of them challenges our conventional understanding of "reality".

1.3 Notes

1. Alexander Herzen, *My Past and Thoughts*, University of California Press, 1999.

2. Quoted in: John Polkinghorne, *The Faith of a Physicist*, Fortress Press, 1996.

3. Steven J. Gould, *Rocks of Ages: Science and Religion in the Fullness of Life*, Random House, 1998.

4. Robert Jastrow, *God and the Astronomers*, Readers Library, 2000.

5. Ian Barbour, *When Science Meets Religion*, Harper Collins, 2000.

6. Robert J. Russell, William R. Stoeger and George V. Coyne (Eds.), *Physics, Philosophy and Theology*, Vatican Observatory Publications, 1988.

7. C. Stephen Layman, "Faith Has Its Reasons," in *God and the Philosophers*, Thomas V. Morris (Ed.), Oxford University Press, 1994.

8. Quoted in: Robert Root-Bernstein, "The Sciences and Arts Share a Common Creative Aesthetic," in *Aesthetics and Science*, Alfred Tauber (Ed.), Kluwer Academic Publishers, 1997.

9. Gerald Holton, *Science and Anti-Science*, Harvard University Press, 1997.

10. Quoted in: Russell *et. al.* (Eds.), *Physics, Philosophy and Theology*.

11. Wallace Clift, *Jung and Christianity*, Crossroads, 1986.

12. This example represents a variant of the one described in David Bartholomew, *Uncertain Belief*, Oxford University Press, 2000.

13. William P. Alston, "A Philosopher's Way Back to Faith," in *God and the Philosophers*, Thomas V. Morris (Ed.), Oxford University Press, 1994.

14. Quoted in: Quentin Lauer, *G. K. Chesterton, Philosopher Without a Portfolio*, Fordham University Press, 1992.

15. Michael J. Buckley, *The Catholic University as Promise and Project*, Georgetown University Press, 1998.

16. Jane Austen, *Pride and Prejudice*, Penguin Books, 1994.

17. Michael J. Himes, "Finding God in All Things: A Sacramental Worldview and Its Effects," in *As Leaven in the World: Catholic Perspectives on Faith, Vocation and the Intellectual Life*, Thomas Landy (Ed.), Rowman and Littlefield, 2001.

18. This categorization was proposed by M. Scott Peck, in *The Different Drum: Community Making and Peace*, Simon and Schuster, 1987.

19. Peck, *The Different Drum: Community Making and Peace*.

20. George Smith is an atheist philosopher, and author of *Atheism: The Case Against God*, Prometheus Books, 1989.

21. Quoted in Brian Davies, *An Introduction to the Philosophy of Religion*, Oxford University Press, 1993.

Chapter 2

Chaos, Complexity and Self-Organization

"Chaos is anti-reductionist. This new science makes a strong claim about the world, namely, that when it comes to the most interesting questions, about order and disorder ... and life itself, the whole cannot be explained in terms of the parts. There are fundamental laws about complex systems, but they are new kinds of laws. They are laws of structure and organization and scale, and they simply vanish when you focus on the individual constituents of a complex system - just as the psychology of a lynch mob vanishes when you interview individual participants." *James Gleick* [1]

The term "chaos theory" is a bit of a misnomer, since the phenomena that it deals with only *appear* to be random. In reality, these phenomena exhibit a subtle, high-level regularity that is not easily perceived, and can only be described in mathematical terms. Prior to the discovery of chaos in the early 1960s, physical processes were routinely classified as *either* deterministic *or* uncertain, but never both at once. The recognition that these two characteristics can coexist in a single system has been one of the great achievements of modern science. We now know that there are systems that inhabit the elusive "middle ground" between order and disorder, where genuine complexity and novelty can emerge.

2.1　What Is Chaos?

The practice of modeling physical systems using differential equations dates back to the 17th century, and the development of Newtonian mechanics. Over the past three hundred years, mathematical descriptions of this type have been successfully applied in a wide variety of disciplines, ranging from physics and engineering to economics and biology. Almost without exception, the underlying equations allowed scientists to accurately predict the dynamics of such systems, and explain the observed experimental data. As a result, there was no reason to suspect that a purely deterministic model could possibly be associated with uncertainties of any kind.

Systems where chaos has been observed are also described in terms of differential equations, which are as simple and precisely defined as those of classical mechanics. And yet, it turns out that these systems possess some very unusual properties, which cannot be anticipated from the structure of the mathematical models that govern their dynamics.

Perhaps the easiest way to describe these properties is to contrast chaos to the behavior of a classical system such as the pendulum shown in Fig. 2.1.

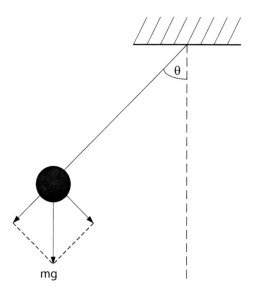

Fig. 2.1. Diagram of a pendulum.

A pendulum has all the characteristics that one would expect to see in a deterministic mechanical system. It's dynamics can be described by two simple first order differential equations, and its trajectory is completely predictable once we provide the initial angle and velocity (the so-called *initial conditions*). The typical behavior of a pendulum is illustrated in Figs. 2.2 and 2.3, which show how the angle $\theta(t)$ and velocity $\omega(t)$ evolve over time. It is obvious that both variables exhibit a regular oscillatory pattern, and that the system stops moving after a sufficiently long time (ultimately settling down into an *equilibrium state*). We should also point out that these plots will *not* change in any meaningful way if the initial conditions are slightly altered. Such insensitivity to small variations ensures that repeated physical experiments with this system will always yield consistent results, although we can never perfectly replicate all the conditions.

In the analysis of dynamic systems such as a pendulum, it is common practice to supplement the graphs shown in Figs. 2.2 and 2.3 with a plot of the velocity ω as a function of angle θ. If we were to do that, we would see that all solutions ultimately converge to the equilibrium point $(\theta^e, \omega^e) = (0,0)$, regardless of where they start from. In this case the equilibrium acts as a sort of "magnet" that

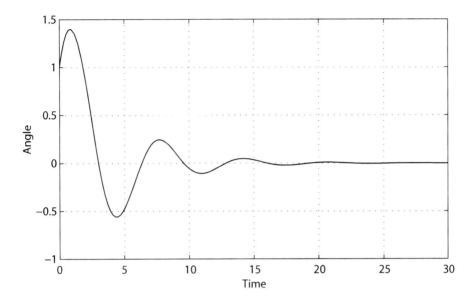

Fig. 2.2. The angle of the pendulum.

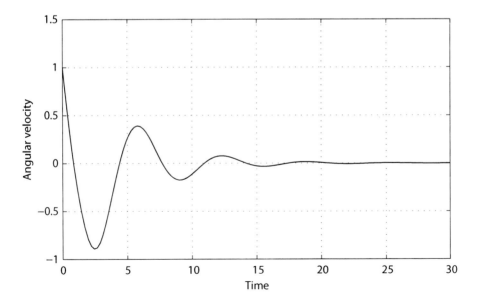

Fig. 2.3. The angular velocity of a pendulum.

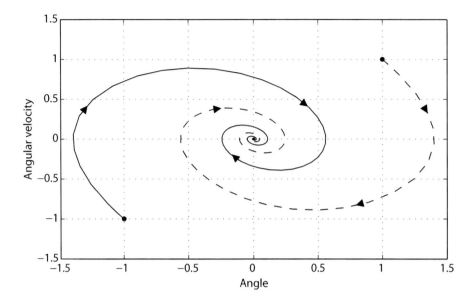

Fig. 2.4. Two trajectories for a pendulum.

"draws in" different trajectories, which is why we refer to it as an *attractor*. This property is illustrated in Fig. 2.4, where we show two different trajectories of a pendulum - one starting out from initial condition $(\theta_0, \omega_0) = (1, 1)$, and the other from $(\theta_0, \omega_0) = (-1, -1)$. Both of them obviously converge to point $(0,0)$.

There are, of course, many other physical systems in which equilibria act as attractors. Figure 2.5 (which represents a predator-prey model in theoretical biology) shows that in some cases there are actually several such equilibria, each one "drawing in" a different set of trajectories. Note that in this particular example some solutions *diverge*, and have nothing to do with the pair of attractors (these would be the two trajectories originating in the lower right-hand corner of the diagram).

What can we conclude from this brief discussion? If we take the pendulum to be a typical representative of deterministic systems, we can say that they are expected to have the following characteristic features:

1) Their dynamic behavior should be *completely predictable* given the equations and a set of initial conditions.

2) Small variations in the initial conditions should have very little effect on the system trajectories.

3) The *attractors* in such systems should be relatively simple geometric forms (equilibria, for example, are *points*).

It turns out that chaotic systems do not possess *any* of these properties, although they are deterministic systems in their own right. Their dynamic behavior is *unpredictable*, they are *hypersensitive* to changes in initial conditions, and their attractors are *geometrically complex* figures. This is completely unex-

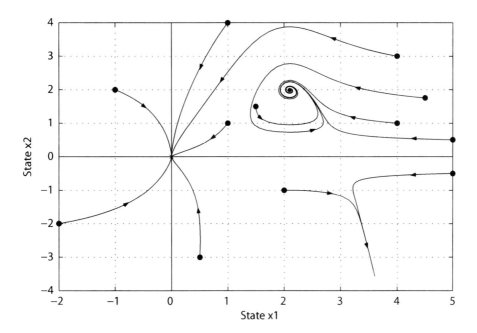

Fig. 2.5. A system with multiple attractors.

pected, since the equations that describe them are precisely defined and contain no random elements.

To illustrate this point, in Fig. 2.6 we show the temporal evolution of one such system, which is used to model the behavior of atmospheric processes. The trajectory depicted in this graph gives no indication whatsoever of any regularity or predictability.

In addition to having random-looking trajectories like the one in Fig. 2.6, chaotic systems can also exhibit a property known as *intermittency*, which allows for sudden and completely unexpected aperiodic bursts that interrupt long periods of orderly behavior. In Fig. 2.7 we show an example of such a burst, which is associated with the so-called *logistic map* (this is a simple equation that has been used for studying a wide range of phenomena, including financial markets, chemical reactions, moth colonies and even human physiology).

Perhaps the most striking aspect of intermittent dynamics is the fact that the aperiodic episodes are essentially unrepeatable, and cannot be predicted from the equations that describe the process. The possibility that a deterministic system could behave in such a way brings into question some widely held assumptions about natural phenomena and the laws that govern them. Indeed, it is conceivable that entire generations of observers could register only the regular dynamic pattern, with no idea that other forms of behavior are even possible. From an empirical standpoint, what we have here are "laws" that can occasionally be "broken", although the underlying equations are no less precise than those of Newtonian physics.

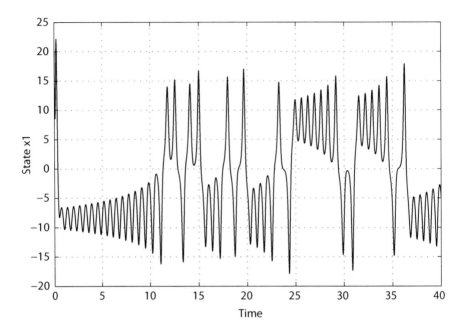

Fig. 2.6. The trajectory of a chaotic system.

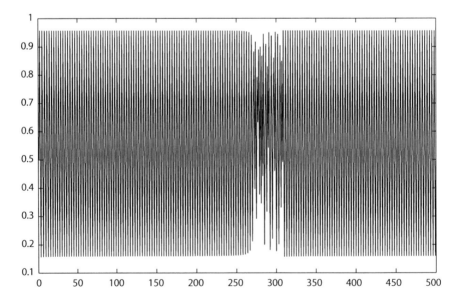

Fig. 2.7. Intermittent behavior of a chaotic system.

To complicate matters even further, we should note that chaotic systems are extraordinarily sensitive to changes in external parameters. Meteorologist Edward Lorenz first observed this phenomenon in 1961, while studying a simple model for weather prediction. In repeating one of his computer simulations, he happened to round-off one of the initial conditions at the fourth decimal place, fully expecting that this would have no significant effect on the calculations. To his surprise, Lorenz found that the new result was completely different from the old one. He subsequently named this phenomenon the "butterfly effect," a term that underscores the possibility that something as insignificant as the movement of a butterfly's wings could ultimately affect global weather patterns.

The extent of the sensitivity exhibited by chaotic systems is illustrated in Fig. 2.8, where we compare two trajectories whose initial values differ by one millionth of one percent (for better clarity, the trajectories are represented by lines of different thickness). It is readily observed that the two solutions are identical at first, but bear no resemblance to each other after a sufficiently long time. This is precisely the kind of dynamic behavior that characterizes atmospheric phenomena, which explains why long term weather forecasts tend to be inaccurate.

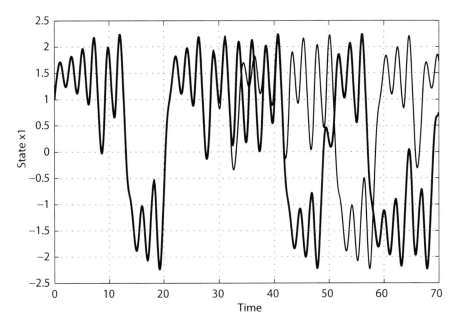

Fig. 2.8. Evolution of two initially close solutions.

In the face of such inherent uncertainty, can we say anything concrete about the behavior of chaotic systems? It turns out that we can. While we cannot precisely anticipate what such a system will do, we *can* establish certain limits regarding what is possible. These limits may be difficult to see if we plot the temporal evolution of the variables, but a graph like the one shown in Fig. 2.4 reveals that there is some order in chaos. It becomes clear, for instance, that such systems *do* have attractors, but with geometric properties that are far more complex than the ones we have encountered so far. An example of such a *strange attractor* is shown in Fig. 2.9 (this particular attractor corresponds to a nonlinear electric circuit). It is interesting to note that although various trajectories traverse this unusual geometric form in unpredictable ways, the difference between them is necessarily *bounded* (since they are all confined to the attractor). This places a strict limit on the range of possible responses, although we cannot specify the exact path that the system will follow.

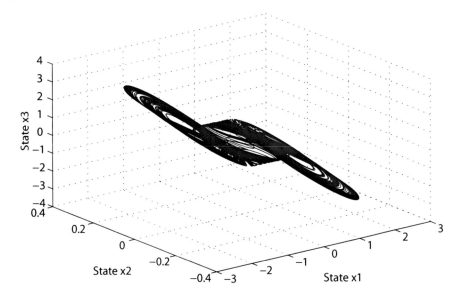

Fig. 2.9. Example of a strange attractor.

The fact that deterministic physical models can sometimes give rise to unpredictable dynamic behavior has important consequences for our understanding of science and its methods of investigation. To begin with, we must allow for the possibility that certain types of complex systems cannot be analyzed using standard experimental techniques. To see why this is so, it suffices to recall that classical physics implicitly assumes that all experiments are *repeatable*. This cannot be guaranteed in the case of chaotic phenomena, where even the smallest variation in external conditions can completely alter the outcome.

The hypersensitivity associated with chaos points to a second difficulty, which is related to the modeling of complex physical processes. If we acknowledge that there are systems whose sensitivity to external influences is infinite (as chaos theory suggests), then there will be instances where it is virtually impossible to distinguish between important and unimportant parameters in the model. In such cases we would have to take the *entire* environment into account, down to the last atom in the most remote corner of the universe. This is clearly a task that exceeds our computational capabilities (and even our imagination).

If standard techniques such as laboratory experiments and numerical simulations fail to provide adequate information about complex phenomena, what other options do we have for describing them? At present, there is no clear answer to this question. The only thing that we can state with reasonable confidence is that the study of complex systems will probably require the development of new kinds of laws, which may not be as "binding" as the ones we are used to. Needless to say, this possibility has interesting theological implications, which we will explore in Chapter 10.

2.2 Strange Attractors and Fractals

Why do we use the attribute "strange" to describe chaotic attractors? The answer to this question appears to be obvious - one simply needs to contrast their highly unusual shapes with the shapes of more "conventional" attractors. There is, however, another reason for using this term, which has to do with the mathematical characterization of such objects. It turns out that chaotic attractors are *fractals*, which are considerably more complicated geometric forms than lines, surfaces or volumes. This is yet another feature that distinguishes chaotic systems from their "well-behaved" counterparts.

Fractals have non-integer dimensions (such as 0.6 or 1.7, for example), which may sound completely counterintuitive to those who are used to classifying objects as one, two or three dimensional. And yet, such forms exist all around us. Modern geometry allows us to assign dimensions to irregular shapes such as coastlines, clouds, forests, and *none* of these turn out to be integers.

Fractal structures can also be generated by computer programs, which typically amount to a set of simple recursive rules. The patterns resulting from these computational procedures can be very intricate and aesthetically pleasing, as can be seen from Fig. 2.10. What is particularly striking about such objects is the fact that some form of order emerges on *every* level of magnification. Indeed, if we were to "zoom" into any one of the spirals in Fig. 2.10, we would discover a set of new and equally fascinating geometric shapes.

To better understand what fractal dimensions mean and how they can be calculated, we first need a definition of dimensionality which is broader than the traditional Euclidean one. While there are a number of such definitions, in the following we will focus on one which is known as *capacitive dimension*. In describing this notion, it is helpful to imagine a sheet of paper that has randomly distributed points and lines on it, as well as areas of empty space (a situation that is not altogether uncommon in modern abstract painting). We will now divide the paper into 'squares' with sides of length ε, where ε is initially a number of our choice. If $N(\varepsilon)$ denotes the number of squares that contain points and lines, we can introduce a quantity $C(\varepsilon)$, which is defined as

$$C(\varepsilon) = \frac{\ln N(\varepsilon)}{\ln(1/\varepsilon)} \tag{2.1}$$

where the symbol ln denotes the natural logarithm. By systematically decreasing the value of ε (while computing the corresponding $C(\varepsilon)$ in each step), we will

Fig. 2.10. Example of a fractal.

ultimately reach the limiting value

$$d_C = \lim_{\varepsilon \to 0} C(\varepsilon) \qquad (2.2)$$

which is known as the capacitive dimension of our set.

The above definition is remarkably general, and can be applied to virtually any geometric object. When used to describe ordinary lines and surfaces, it produces values $d_C = 1$ and $d_C = 2$, respectively, which clearly conforms to our usual understanding of dimensionality. On the other hand, this technique can also be applied to all kinds of irregular shapes, ranging from those found in nature to the abstract paintings of Jackson Pollock. What is particularly interesting about Pollock's work is his signature "drip and splash" technique, which involved an apparently random splattering of paint onto a canvas that was placed on the floor of his studio. It turned out, however, that the results were not random at all. A careful analysis showed that his paintings had a definite fractal dimension, which increased as he matured as an artist.[2]

The following simple example further illustrates how fractals differ from ordinary geometric objects, and how their dimension can be computed in a systematic manner.

Example 2.1. Let us consider the procedure outlined in Fig. 2.11, which produces a rather unusual geometric object. In the first step of this algorithm, a line of unit length is partitioned into three equal segments. The middle segment is then replaced by an equilateral triangle, whose bottom side is removed.

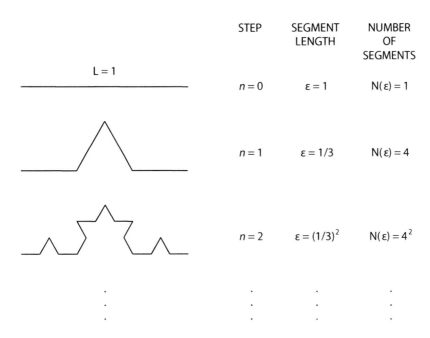

	STEP	SEGMENT LENGTH	NUMBER OF SEGMENTS
$L = 1$	$n = 0$	$\varepsilon = 1$	$N(\varepsilon) = 1$
	$n = 1$	$\varepsilon = 1/3$	$N(\varepsilon) = 4$
	$n = 2$	$\varepsilon = (1/3)^2$	$N(\varepsilon) = 4^2$

Fig. 2.11. Construction of the Koch curve.

If this simple operation is recursively repeated on each segment, after n steps we will obtain a rather "jagged" looking collection of $N(\varepsilon) = 4^n$ line segments, each of length $\varepsilon = (1/3)^n$. When $n \to \infty$, this object turns into what is known as a *Koch curve*, whose capacitive dimension is

$$d_C = \lim_{\varepsilon \to 0} \left[\frac{\ln N(\varepsilon)}{\ln(1/\varepsilon)} \right] = \lim_{n \to \infty} \left[\frac{\ln(4^n)}{\ln(3^n)} \right] = 1.261 \qquad (2.3)$$

Since the value of d_C lies between 1 and 2, it is fair to say that the Koch curve is in some sense more complex than a line, but less complex than a surface. This is not something that we encounter in dealing with "ordinary" geometric forms, whose dimension always take on integer values.

It is interesting to note that the Koch curve actually has an "invisible" microscopic structure whose cumulative length is *infinite*. To get a sense for how something like this is possible, imagine that we have a measurement device with an exceptionally high resolution, which can distinguish segments that are as small as 10^{-15} meters (this is roughly the diameter of an atomic nucleus). If our initial line happened to be 1 meter long, such a device would presumably allow us to precisely record how the length of the object evolves during the first 32 steps of the construction process (since $(1/3)^{32} = 0.54 \cdot 10^{-15}$).

The maximal length that we would be able to measure in this way is

$$L_{\max} = (4/3)^{32} = 9.95 \cdot 10^3 \tag{2.4}$$

which is slightly less than 10 kilometers. Beyond that limit, changes in the segments simply become too small for detection. We do know, on the other hand, that the length of the Koch curve grows as

$$L(n) = (4/3)^n \tag{2.5}$$

and therefore tends to infinity as n increases. This, however, is something that our measurement instruments *can not* detect. From our perspective, the curve "freezes" at a point that is defined by the resolution of the measurement apparatus, and appears to have a finite length. This explains how such an object can be "neatly packed" into a unit square, despite the fact that $L \to \infty$.

2.3   Complexity and Self-Organization

The phenomenon of chaos is intimately related to the notions of *complexity* and *self-organization*, and this connection is worth examining in some detail. Given that complexity is a notoriously ambiguous word, it makes sense to begin our discussion with an appropriate definition. This is by no means easy, since there are a number of competing interpretations in current use. In computer science, for example, the term "complexity" is usually associated with the length of the shortest program that can produce a particular data set. If we adopt that definition, a number such as π could be viewed as relatively "simple," since there are explicit formulas that allow us to compute its decimal digits. Because such formulas exist, the *same* short program could be used to produce the first 5 or the first 100,000 decimal of this number.

In contrast to π, there is an abundance of real numbers whose decimal digits exhibit *no* pattern whatsoever. Such numbers are said to have high *algorithmic information complexity*, since each decimal digit needs to be described *separately*. This means (among other things) that the program needed to produce the first 5 decimal digits would be fairly short, but the one needed for the first 100,000 decimals would be extremely long. The program length would obviously continue to grow if we were to increase the desired number of decimals, with no upper limit in sight.

While the above definition appears to be perfectly acceptable in information theory, it clearly deviates from what we normally mean when we use the attribute "complex."

> "Compare a play by Shakespeare with the typical product, of equal length, of the proverbial ape at the typewriter, who types every letter with equal probability. The algorithmic information complexity, or algorithmic randomness, of the latter is overwhelmingly likely to be much greater than that of the former. But it is absurd to say that

the ape has produced something more complex than the work of Shakespeare." *Murray Gell-Mann* [3]

To this we might add the observation that random data can often be represented in compact form, despite the fact that it cannot be reproduced by a simple program. A typical example of this sort is so-called "white noise," which is completely random but can nevertheless be described in terms of only two pieces of information - the mean and the standard deviation. With that in mind, it seems reasonable to treat *both* strictly regular *and* strictly irregular data as essentially "simple," while viewing data that lies somewhere in between as "complex." Complexity understood in this way can be directly associated with phenomena such as chaos, which have both orderly and unpredictable features.

The same combination of order and disorder that characterizes chaotic behavior also plays a crucial role in the process of *self-organization*. Recent research suggests that complex systems (both of the animate and inanimate variety) seem to "prefer" operating at the edge of chaos, where the range of possible configurations is virtually unlimited. Such "openness" to change, combined with an extraordinary sensitivity to external stimuli, is an essential prerequisite for the emergence of novelty in nature.

It goes without saying, of course, that novelty alone doesn't automatically imply an increase in complexity (or even functionality, for that matter). Evolutionary biology makes this point in no uncertain terms. Nevertheless, the sheer number of possible outcomes that exist at the "edge of chaos" makes it very likely that some of them will ultimately lead to higher forms of organization. More often than not, such "advanced" forms arise spontaneously, as a result of simple *local* interactions between autonomous agents. These interactions tend to occur without any form of global coordination, which is why they are commonly associated with the process of self-organization.

Phenomena of this sort are usually characterized by one or more threshold values, which separate the domains of "ordinary" and "critical" behavior. The term "critical" implies (among other things) that small disturbances in the system can have unpredictable consequences. As an illustration of this property, imagine a large object that is being pushed across a smooth surface (such as ice, for example). In that case, energy will be dissipated at the same rate at which it is added to the system, and the motion of the object will be *continuous*. If, on the other hand, the surface happens to be rough, energy could be added to the system for quite a while without causing any motion at all. However, when a threshold value is reached, the accumulated energy dissipates through an abrupt, *discontinuous* movement. States in which a force is applied but the object remains motionless are extraordinarily sensitive to external disturbances. When the system is in such a state (which corresponds to the "edge of chaos"), a small increase in the applied force can bring about almost any kind of response - a small movement, a large one, or even no movement at all. It is this combination of unpredictability and hypersensitivity that creates the necessary conditions for novelty and increased complexity.

Among the many instances of spontaneous self-organization in nature, the ones that arise in biology are perhaps the most striking. We do not have to look too far up the evolutionary chain to find evidence of such processes. Certain types of molds, for example, produce amoeba-like cells that tend to "cluster together" when the food supply becomes scarce. In doing so, they follow changes in the concentration of a chemical substance know as acrasin, which is secreted by the cells when there is a shortage of food. Interestingly, each cell that encounters such a marker begins to produce its own "acrasin trail," thus amplifying the local chemical signal. The end result of this process is an aggregation of cells that resembles a rather large slug-like creature, which is a more complex and qualitatively different organizational form.

An important aspect of this example is the complete absence of any global coordination among the cells. The secretion and detection of chemical signals are essentially forms of *local* interaction, which ultimately gives rise to a higher-level structure. Something similar happens when termites construct their mounds, or when ants create swarms (whose length can be up to several hundred meters). In both cases, the individual ants have no way of knowing that they are part of a larger unit - they simply follow a chemical trail left by other ants in their immediate vicinity.

> "There is no one ant that is calling the shots, picking from among all the other ants which one is going to get to do its thing. Rather, each ant has a very restricted set of behaviors. ... When one takes these behaviors on aggregate, the whole collection of ants exhibits a behavior, at the level of the colony itself, which is close to being intelligent. ... A collective pattern takes over the population, endowing the whole with models of behavior far beyond the simple sum of the behaviors of its constituent individuals." *Christopher Langton* [4]

It is interesting to note that the organizational patterns exhibited by both the molds and the ants have been successfully reproduced by computer-aided simulation.[5] In all cases, the programs were based on very simple forms of local interaction between individual agents, but the overall system exhibited complex dynamic behavior. Computer models based on simple local rules were also found to be effective in the study of gene regulation. Pioneering work in this field was performed by biologist Stuart Kauffman, who introduced *random Boolean networks* as a theoretical tool for modeling this process.[6] A Boolean network can be viewed as a collection of N interconnected nodes (interpreted in this case as genes), which are capable of activating or deactivating each other according to a set of preassigned rules. Despite its apparent simplicity, this mathematical framework proved to be very useful in describing biological systems, and has produced predictions that are remarkably consistent with observed patterns of cell differentiation in living organisms.

To get a better sense for how Boolean networks operate, in Fig. 2.12 we provide a simple schematic illustration of such a system. Note that each node in Fig. 2.12 has both *incoming* and *outgoing* edges, whose directions reflect the flow of information in the network. The information exchange between the nodes is strictly *local*, and there is no global coordination (which is typical for systems that are capable of self-organization).

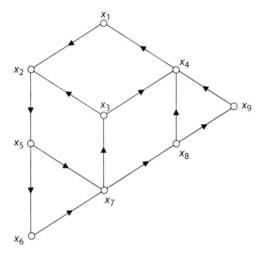

Fig. 2.12. Example of a Boolean network.

The nodes in a Boolean network can be viewed as *autonomous agents*, which act independently in response to external stimuli. It is usually assumed that the state of each node evolves discretely over a series of time steps. In mathematical terms, this means that the next action of node i can be described as

$$x_i(k+1) = F_i(x(k)) \tag{2.6}$$

where vector $x(k) = [x_1(k) \dots x_N(k)]$ represents the current states of all neighboring nodes that affect it. It is important to recognize in this context that functions F_i $(i = 1, 2, \dots, N)$ are Boolean, which means that the only values they can produce are 0 or 1. It is easily verified that any such function can be defined in the form of a look-up table, which is convenient from the standpoint of computer-aided simulation.

Assuming that each of the N nodes has exactly K incoming edges, there are many possible ways for assembling a Boolean network (each such configuration represents a "realization" of the NK network). In addition to selecting an interconnection pattern, one can also assign a different function F_i to each node and choose a set of initial conditions. When all these options are taken into account, the number of possibilities can reach staggering proportions. For that reason, the study of Boolean networks normally involves a *random* sampling of different configurations (hence the name "random Boolean networks").

In order to see how connectivity affects the process of gene regulation, Kauffman studied the average number and length of independent periodic patterns (so-called *limit cycles*) that arise in an ensemble of randomly chosen NK networks. His experiments showed that when each gene interacts with only *one* neighbor (*i.e.* when $K = 1$), the system dynamics become rather uninteresting, with a tendency of converging to a fixed set. On the other hand, if genes were allowed to exchange information with *all* other genes (*i.e.* when $K = N$), the patterns were found to be extraordinarily complicated and hypersensitive to perturbations. In most such cases, the overall system behavior showed no signs of regularity. At the transition between these two rather extreme regimes, Kauffman identified a region of "critical" behavior, which is characterized by a mixture of order and disorder. He referred to this region as the "edge of chaos," and suggested that such complex dynamics provide a natural mechanism for the process of self-organization in biological systems. [7]

The patterns associated with Boolean networks that operate at the "edge of chaos" highlight the importance of local interactions, and point to *complexity* as a necessary condition for information processing on the cellular level. Indeed, when the system is "static" (as in the case of $K = 1$), information can be stored but cannot be manipulated in any meaningful way. On the other hand, systems that change very easily (such as those with $K = N$) are capable of processing information, but are too dynamically active to allow for any kind of memory. The complex behavior that occurs for the critical value $K = 2$ seems to provide just the right balance of order and uncertainty that enables the nodes to transmit, modify and retain information (which is what genes are presumably supposed to do). [8]

What can we conclude from this discussion? Based on the examples considered above, it would appear that new and increasingly complex structures can arise spontaneously in nature, simply as a result of local interactions between autonomous agents. This type of self-organization does not require a "master plan," or any other form of global control.

The fact that such processes are closely associated with chaos further indicates that a mixture of order and uncertainty is essential for the emergence of novelty in nature. Indeed, if all physical processes were as orderly and predictable as the movement of a pendulum, it is highly unlikely that life as we know it would ever have developed. Fortunately for us, the universe seems to be a far more interesting place than physicists of the 18th and 19th centuries had anticipated. What we have learned about chaos and complexity in recent years suggests that nature is still a "work in progress", whose future states are by no means predetermined. From that perspective, it is perhaps best to view our physical environment as a vast array of possibilities, not all of which will be realized. We will see in Chapter 10 that such a view has important consequences for the theological understanding of evolution and the origins of the universe.

2.4   Notes

1. Quoted in: Steven Weinberg, *Dreams of a Final Theory*, Pantheon, 1992.

2. For further details, see: Richard Taylor, "Fractal Expressionism - Where Art Meets Science," in *Art and Complexity*, J. Casti and A. Karlqvist (Eds.), Elsevier Science, 2003.

3. Murray Gell-Mann, "Regularities and Randomness: Evolving Schemata in Science and in the Arts," in Casti and Karlqvist (Eds.), *Art and Complexity*.

4. Quoted in: Charles Birch, "Neo-Darwinism, Self-Organization and Divine Action in Evolution," in *Evolutionary and Molecular Biology*, R. J. Russell, W. R. Stoeger and F. J. Ayala (Eds.), Vatican Observatory Publications, 1998.

5. See *e.g.* Mitchell Resnick, "Learning About Life," in *Artificial Life: An Overview*, Christopher Langton (Ed.), MIT Press, 1995. Another interesting article on this subject is: Nigel Franks, "Army Ants: A Collective Intelligence," *Scientific American*, **77**, pp. 139-145.

6. Stuart Kauffman, *Origins of Order: Self Organization and Selection in Evolution*, Oxford University Press, 1993.

7. Ibid.

8. An interpretation of Boolean networks which links complexity and information processing can be found in: Gary Flake, *The Computational Beauty of Nature*, MIT Press, 2000.

Chapter 3

Metamathematics

"Gödel's Incompleteness Theorem, Church's Undecidability Theorem, Turing's Halting Problem - all have the flavor of some ancient fairy tale which warns you that 'To seek self-knowledge is to embark on a journey which ... will always be incomplete, cannot be charted on a map, will never halt, cannot be described.'" *Douglas Hofstadter* [1]

In conventional mathematics we generally start out with established axioms and theorems, and utilize precisely defined rules and procedures to prove new results. The proofs themselves are usually combinations of symbols and words that explain the logic behind this process. As the prefix "meta" suggests, metamathematics goes beyond this traditional approach, and is interested in the *structure* of proofs.

In analyzing the way in which theorems are derived from axioms, it is often useful to represent all statements and arguments that constitute a proof in terms of numbers. Such a "translation" is sometimes referred to as the *arithmetization* of metamathematics, which can assume a variety of different forms. To get a sense for what this entails, consider the fact that the word "proof" can be encoded as $\{16, 18, 15, 15, 6\}$ by simply replacing each letter with the number that corresponds to its position in the alphabet. A more sophisticated approach would be to systematically assign a *single* number to each finite string of symbols. Given such a mapping, one could potentially translate all statements of a logical system into statements about integers. This kind of abstract representation lies at the heart of metamathematics, and is closely related to Gödel's famous Incompleteness Theorem.

The material presented in this chapter is designed to provide some insight into Gödel's fundamental result, whose implications for mathematics have been momentous. The concepts and ideas introduced in this chapter point to certain inherent limitations of human logic, which will subsequently be invoked in our discussion of the relationship between science and religion (see Chapters 9 and 10).

3.1   What Is a Formal System?

The 'axiomatic' approach to mathematics dates back to ancient Greece and the development of geometry. The starting point of the Greeks was quite simple - they argued that geometry can be based on a number of self-evident propositions that are accepted *without proof*. These propositions are known as *axioms*, and can subsequently be used to derive all other propositions as *theorems*, according to some preset rules.

Around 300 B.C., Euclid set out to systematize all existing knowledge about geometry in a seminal book entitled *The Elements*. He began by formulating the following four axioms, which represent rather obvious statements reflecting our everyday experience.

AXIOM 1. A straight line segment can be drawn joining any two points.

AXIOM 2. Any straight line segment can be extended indefinitely in a straight line.

AXIOM 3. Given any straight line segment, a circle can be drawn having the segment as radius and one end point as center.

AXIOM 4. All right angles are congruent.

From these axioms, the first 28 theorems of geometry were derived directly (they constitute what is known as *absolute geometry*). But there was a catch ... Despite all his efforts, Euclid was unable to prove a proposition that can be paraphrased as follows:

PROPOSITION 5. Consider a line l and a point a that does not belong to it. Then, there is one and only one line that passes through a and does not intersect l.

Since this statement appeared to be intuitively true, Euclid decided to adopt it as the fifth axiom in his system. All subsequent theorems that followed from this assumption constitute what we now refer to as Euclidean geometry.

There is no doubt that Euclidean geometry developed into a logically sound system. Since its axioms reflected obvious truths, there was no danger that two theorems would ever contradict each other (a property known as *internal consistency*). Over the years, the practical applicability of Euclidean geometry added to its appeal, and solidified its status as a cornerstone of mathematics. And yet, many mathematicians remained troubled by the fifth axiom. Part of the problem stemmed from the belief that human intuition is limited to *finite* objects, and cannot be used to anticipate what may happen at infinity (which is precisely what Proposition 5 required). As a result, the two millennia that followed Euclid's death witnessed countless attempts to close the loophole, and derive Proposition 5 from the original four axioms. These efforts proved to be unsuccessful, and by the end of the 19th century it was finally established that such a derivation is impossible.

The 'undecidability' of Proposition 5 turned out to be an enormously important result. Among other things, it was now possible to argue (with full

logical justification) that a statement such as the following one can be used as
an alternative to Proposition 5:

PROPOSITION 5A. Consider a line l and a point a that does not belong to
it. Then, there is *no line* that passes through a and does not intersect l.

The introduction of Proposition 5A as an axiom seems rather bizarre and
thoroughly counterintuitive. How can geometry possibly be based on a statement
that clearly contradicts our everyday experience? Wouldn't this be comparable
to denying gravity or the rotation of the earth? It certainly would, if we were to
adhere to our intuitive notions of a point, line and other basic geometric forms.
But what if we were to define a 'point' as a *pair* (a, a_1) of diametrically opposite
points on a sphere, and a 'line' l as a *great circle*? In that case, Proposition 5A
becomes perfectly reasonable. To see why this is so, consider a line l and a point
(a, a_1) that does not belong to it (as shown in Fig. 3.1). It is easily verified that
any line l_1 passing through (a, a_1) must intersect line l.

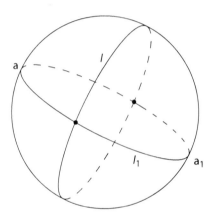

Fig. 3.1. Generalized notions of a point and line.

One could argue at this point that axioms such as Proposition 5A may be
technically acceptable, but are so abstract that the resulting systems would be
completely detached from reality. It turns out, however, that this is not the case.
Non-Euclidean geometries have actually found numerous applications, which
range from economics to Einstein's theory of general relativity (see Chapter 5 for
more details). Developments such as these have lead mathematicians to conclude
that axioms needn't be self-evident truths, or even facts that correspond to our
experience. Axioms can actually be pretty much anything, provided that the
ensuing theorems do not result in a contradiction.

The recognition that axioms can be chosen in a flexible manner opened the
door to the formalization of mathematics. Almost without exception, theo-
reticians came to believe that all mathematical reasoning could be ultimately
represented as a *formal system*, in which the axioms and theorems are strings of
completely meaningless symbols. Thinking along these lines, the famous philosopher and logician Bertrand Russell came to the conclusion that:

> "Mathematics may be defined as the subject in which we never know
> what we are talking about, nor whether what we are saying is true."
> *Bertrand Russell* [2]

If we agree with Russell that mathematical concepts needn't have any meaning, then the only things that matter are the choice of symbols and the rules by which we manipulate them. Specifying these rules is usually a fairly easy thing to do. It turns out, for example, that describing a formal system requires only three pieces of information:

THE RULES OF FORMATION, which determine whether or not a string of symbols is a legitimate member of the system.

THE AXIOMS, which are "truths" that are accepted without proof.

THE RULES OF INFERENCE, which provide the basis for generating new theorems from axioms and already existing theorems.

The following example shows what a simple formal system might look like. [3]

Example 3.1. (The PQ system). Consider a formal system in which there are only three types of symbols: letters P, Q, and strings of hyphens. The system is defined by the rules and axioms described below.

Rule of Formation. Any string of the type

$$x \, P \, y \, Q \, z \tag{3.1}$$

where x, y and z represent sequences of one or more hyphens is a member of this system (we will refer to it as a *well-formed string*).

Axioms. Every string of the type

$$x \, P - Q \, x- \tag{3.2}$$

where x is a sequence of hyphens is an axiom.

Rule of Inference. If

$$x \, P \, y \, Q \, z \tag{3.3}$$

is a theorem, then

$$x \, P \, y - Q \, z- \tag{3.4}$$

is also a theorem.

In analyzing the properties of this system, we should first recognize that certain combinations of P, Q and hyphens are *not* allowed in this system. For example,

$$- - - P - - Q - - - \tag{3.5}$$

is a well-formed string, but

$$- - -PQ - --$$ (3.6)

or

$$-P - Q - P$$ (3.7)

are *not*, since they violate the rule of formation. We should also point out the theorems of the PQ system can be automatically generated from the axioms by applying the rule of inference. Indeed, if we were to start with axiom

$$- - P - Q - --$$ (3.8)

the theorem

$$- - P - -Q - - - -$$ (3.9)

would follow directly. Applying the rule of inference again, we would get

$$- - P - - - Q - - - --$$ (3.10)

and so on. It would actually be quite straightforward to program a computer to generate theorems in this manner. Note, however, that in this case such a program would never terminate, since there are infinitely many axioms and boundless possibilities to produce theorems from them.

The two most important questions that we can ask about any formal system are whether it is *internally consistent,* and whether it is *complete.* What exactly does that mean? For our present purposes, it will suffice to simply describe these concepts, without resorting to precise definitions. We will say that a formal system is internally consistent if its rules of inference ensure that no statement can be *both* true *and* false at the same time. Completeness, on the other hand, will be understood as a property which ensures that every statement in the system can be classified as *either* true *or* false. Note that according to this interpretation, a formal system would be *incomplete* if it contained at least one *undecidable proposition* (*i.e.* a proposition whose "truth status" cannot be established).

Showing that a system is consistent and/or complete can be a formidable task, even in simple cases. If, for example, we were to attempt a proof of consistency based on the rules of inference alone, our only option would be to generate theorems one by one until we find two that stand in logical opposition. In this way, we might be able to establish *inconsistency* in a finite number of steps (if we are very, very fortunate). Consistency, however, can never be verified in this manner, since infinitely many theorems would need to be examined. It follows, therefore, that such a proof requires some form of reasoning that *transcends* the formal system itself.

A possible way to resolve this problem would be to introduce an *interpretation*, which would allow us to "translate" all the statements of the original formal system into statements about some other (preferably more manageable) system. We already saw how something like this might be done in simple cases (see, for example, Table 1.1 in Chapter 1). A somewhat more sophisticated illustration of this process arises in the context of geometry. For centuries it was assumed that the Euclidean axiomatic system must be consistent, given the intuitive nature of its principal propositions. No formal validation of this claim was attempted until the late 1800s, when mathematician David Hilbert proposed a rigorous consistency proof. His idea was remarkably simple, and amounted to associating a pair of (x, y) coordinates with every point in a plane. Such a model allowed him to "translate" every statement of Euclidean geometry into a corresponding algebraic statement, and interpret it accordingly.

The notion of an "interpretation" can be extended well beyond the simple example considered above. A systematic way to do that would be to construct a mapping that assigns a unique integer to each symbol, string and proof of any given formal system. Such an approach seems very promising, since it allows us to "translate" every statement of this system into a corresponding statement about integers, and use the framework of number theory to establish its consistency and completeness. It turns out, however, that this path is a rather treacherous one. As we shall see, it led to some very surprising results, which ultimately brought into question the very foundations of mathematics.

3.2   Gödel's Theorem and Its Consequences

Throughout the early decades of the 20th century, no one seriously doubted the premise that mathematical thinking could be fully captured by formal rules and symbols. This conviction provided the inspiration for the *Principia Mathematica*, a three-volume work by Bertrand Russell and Alfred North Whitehead which was designed to lay the logical foundations of all mathematics, past, present and future. To get a sense for the extraordinary complexity of this endeavor, it will suffice to consider the excerpt from Principia Mathematica shown in Table 3.1, which represents the formal proof that $1 + 1 = 2$.[4]

It was widely believed that such a meticulously constructed logical system could remove all ambiguities and inconsistencies within mathematics. Following the completion of Principia Mathematica, the Platonic vision of an ideal mathematics finally seemed to be within reach. Then came one of the greatest surprises in the history of human thought. In 1931, Austrian mathematician Kurt Gödel published a paper entitled "On Formally Undecidable Propositions of Principia Mathematica and Related Systems."[5] Behind this rather uninspiring heading was a set of results that completely shattered the prevailing view of mathematics.

What exactly did Gödel prove? His main result can be outlined as follows:

Theorem 3.1. (The First Incompleteness Theorem) In every sufficiently complex formal system there are *unprovable propositions*. Even if such a system

$*54 \cdot 42$. $\vdash :: \alpha \in 2. \supset :.\beta \subset \alpha. \; !\beta.\beta \neq \alpha. \equiv .\beta \in \iota``\alpha$

Dem.

$\vdash . *54 \cdot 4. \quad \supset \vdash :: \alpha = \iota`x \cup \iota`y. \supset :.$

$$\beta \subset \alpha. \exists!\beta. \equiv : \beta = \Lambda. \lor .\beta = \iota`x. \lor .\beta = \iota`y. \lor .\beta = \alpha : \exists!\beta :$$

$[*24 \cdot 53 \cdot 56. *51 \cdot 161] \quad \equiv : \beta = \iota`x. \lor .\beta = \iota`y. \lor .\beta = \alpha \tag{1}$

$\vdash . *54 \cdot 25. \text{ Transp. } *52 \cdot 22. \supset \vdash : x \neq y. \supset .\iota`x \cup \iota`y \neq \iota`x.\iota`x \cup \iota`y \neq \iota`y :$

$[*13 \cdot 12] \supset \vdash : \alpha = \iota`x \cup \iota`y.x \neq y. \supset .\alpha \neq \iota`x.\alpha \neq \iota`y \tag{2}$

$\vdash .(1).(2). \supset \vdash :: \alpha = \iota`x \cup \iota`y.x \neq y. \supset :.$

$$\beta \subset \alpha. \exists!\beta.\beta \neq \alpha. \equiv : \beta = \iota`x. \lor .\beta = \iota`y :$$

$[*51 \cdot 235] \qquad\qquad\qquad\qquad \equiv : (\exists z) \cdot z \in \alpha.\beta = \iota`z :$

$[*37 \cdot 6] \qquad\qquad\qquad\qquad\qquad \equiv : \beta \in \iota``\alpha \tag{3}$

$\vdash .(3). *11 \cdot 11 \cdot 35. *54 \cdot 101. \supset \vdash . \text{ Prop}$

$*54 \cdot 43$. $\vdash :.\alpha, \beta \in 1. \supset : \alpha \cap \beta = \Lambda. \equiv .\alpha \cup \beta \in 2$

Dem.

$\vdash . *54 \cdot 26. \supset \vdash :.\alpha = \iota`x.\beta = \iota`y. \supset : \alpha \cup \beta \in 2. \equiv .x \neq y.$

$[*51 \cdot 231] \qquad\qquad\qquad\qquad\qquad \equiv .\iota`x \cap \iota`y = \Lambda.$

$[*13 \cdot 12] \qquad\qquad\qquad\qquad\qquad \equiv .\alpha \cap \beta = \Lambda \tag{1}$

$\vdash .(1). *11 \cdot 11 \cdot 35. \supset$

$\qquad \vdash :.(\exists x, y).\alpha = \iota`x.\beta = \iota`y. \supset : \alpha \cup \beta \in 2. \equiv .\alpha \cap \beta = \Lambda \tag{2}$

$\vdash .(2). *11 \cdot 54. *52 \cdot 1. \supset \vdash . \text{ Prop}$

Table 3.1. An excerpt from Principia Mathematica.

were augmented by an indefinite number of new axioms and rules, there would always be number theoretic truths that are not formally derivable.

Theorem 3.1 basically states that any sufficiently complex formal system will necessarily be *incomplete*, and that nothing can be done to fix the problem. This leads to the inevitable conclusion that mathematics is an inherently imperfect discipline (a fact that many of its practitioners have found exceedingly difficult to accept). As a corollary to his Incompleteness Theorem, Gödel also showed that it is impossible to prove the consistency of a formal system such as the one proposed in Principia Mathematica from within the system itself.

A rigorous discussion of these results requires advanced mathematics, and is clearly beyond the scope of this book. For our purposes, it suffices to say that the most original part of Gödel's proof was the formulation of a statement that was *undecidable* within the system itself. In constructing such a statement, Gödel made extensive use of the possibility of *self-reference*. To gain a better understanding of why this property leads to logical difficulties, let us consider the ancient paradox of Epimenedes, which can be summarized by the statement S: "I am lying." The diagrams in Figs. 3.2 and 3.3 illustrate the logical possibilities, both of which lead to an apparent contradiction.

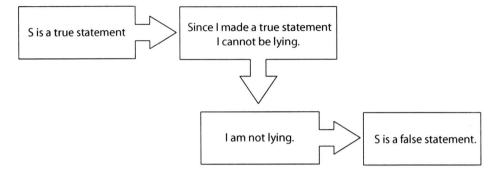

Fig. 3.2. The first possibility.

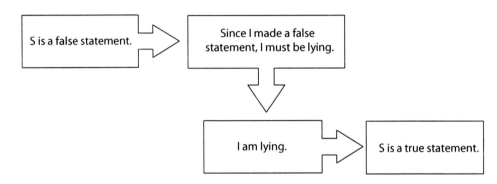

Fig. 3.3. The second possibility.

The source of this paradox lies in the way in which we use the word "true." It obviously appears as a part of sentence S (*i.e.* within the spoken language), but we also use it to say something *about* S. When a word is used in the latter (self-referential) context, we say that it belongs to the *meta-language*. Epimenides and his successors failed to distinguish between these two logical levels, and that caused a great deal of confusion.

In a very real way, Gödel's work opened the door to a new paradigm in mathematics. His Incompleteness Theorem clearly showed that there are many aspects of mathematical research that cannot be reduced to a formal manipulation of symbols. This result effectively eliminated the possibility of automatic proof verification, and ensured that the process of validating theorems would ultimately remain in the hands of qualified experts. Given that experts are fallible human beings, there was no way to guarantee that a consensus could always be achieved, or that subtle errors would not be uncovered at some point in the future.

For those who yearned for certainty, the prospect of errors and formally unprovable propositions was unwelcome news.

> "I wanted certainty in the kind of way in which people want religious faith. I thought that certainty is more likely to be found in math-

ematics than elsewhere. But I discovered that many mathematical demonstrations, which my teachers expected me to accept, were full of fallacies, and that if certainty were indeed discoverable in mathematics, it would be in a new field of mathematics, with more solid foundations than those that had hitherto been thought secure. ... After some twenty years of very arduous toil, I came to the conclusion that there was nothing more that I could do in the way of making mathematical knowledge indubitable." *Bertrand Russell* [6]

It doesn't follow, however, that mathematicians ought to despair because of the apparent shortcomings of their discipline.

"After the initial shock, mathematicians realized that Gödel's theorem, in denying them the possibility of a universal algorithm to settle all questions, gave them instead a guarantee that mathematics can never die." *Freeman Dyson* [7]

The imperfections of mathematics raise an interesting practical question. Namely, in the absence of absolute certainty, what criteria must be met before a proposition can be accepted as *true*? The following example illustrates a fundamental dilemma that arises in this context.

Example 3.2. A real number r is said to be 10-*normal* if every sequence of k consecutive digits in its decimal expansion occurs with a relative frequency of 10^{-k}. This basically means that if we consider a very large number of decimals in r, the number 7 will appear (on the average) once in every 10 digits, the pair 23 once in every 10^2 digits, and so on. Given that there is usually no discernible pattern in the decimal expansions of real numbers, it is reasonable to expect that most of them conform to such a statistical distribution. True enough, in the case of numbers like π (for which more than 10^{13} decimals are known), there is very strong experimental evidence of 10-normality. And yet, neither π nor any other fundamental mathematical constant have been rigorously *proved* to have this property. As a result, most mathematicians are reluctant to accept the 10-normality of π as a *true* statement.

There are, on the other hand, results that have been rigorously established but whose proof is extraordinarily complicated. A typical example is the proof of Fermat's Last Theorem, which states that there are *no* nonzero integers x, y and z such that

$$x^n + y^n = z^n \qquad (3.11)$$

holds when n is an integer larger than 2. This proposition has a very interesting history, dating back to the 17th century. In 1632, French mathematician Pierre de Fermat made the following short note in the margin of a book that he was reading: "I have a truly marvellous proof of this proposition which this margin is too narrow to contain." Unfortunately, many of Fermat's papers were lost after his death, so we will probably never establish how Fermat solved this problem (and if his approach was correct). The only thing we know for sure is that the proof of this conjecture eluded mathematicians for the next 350 years. When Andrew Wiles finally published a proof in 1995, it was so complex that very few experts could grasp it, let alone verify its accuracy. This, however, did not prevent the mathematical community from accepting Wiles' result as a legitimate truth.

Is it fair to say, then, that Fermat's Last Theorem is true, while the 10-normality of π is only a likely possibility? I don't think that there is a simple answer to this question. It could be argued, for example, that it is highly probable that π is a 10-normal number, given the existing experimental evidence. On the other hand, judging by historical experience, it is considerably less likely that Wiles' proof is completely free of errors. After all, mathematics abounds with examples of results where subtle inaccuracies remained undetected for centuries. It is therefore far from clear that rigorous proofs are the *only* acceptable standard of truth when it comes to highly complex problems.

> "Perhaps only 50 or 100 people alive can, given enough time, digest *all* of Andrew Wiles' extraordinarily sophisticated proof of Fermat's Last Theorem. If there is even a 1% chance that each has overlooked some subtle error ... then we must conclude that computational results are in many cases actually more secure than the proof of Fermat's Last Theorem." *Jonathan Borwein* [8]

The above discussion suggests that mathematics as it is really practiced can never be entirely free of ambiguities. In that respect it resembles the natural sciences, where hypotheses are continuously tested, refuted and reformulated, and where propositions are accepted when they are shown to be *sufficiently probable*. Are we then at liberty to conclude that it is just a matter of time before mathematics will "officially" open itself to experimental evidence? Perhaps, but there is certainly no consensus at the present time. What makes this question particularly intriguing is the fact that it directly pertains to the most famous unsolved problem in contemporary mathematics (the so-called Riemann hypothesis). Given that the Clay Institute of Mathematics currently offers one million dollars to anyone who proves this hypothesis, it is probably worth our while to take a closer look at it. Since this example involves some rather sophisticated mathematical concepts, readers who lack the necessary background may want to skip the technical details, and focus only on the philosophical implications.

Example 3.3. (The Riemann Hypothesis). The central issue in the Riemann hypothesis are the zeros of the so-called *zeta function*, $\zeta(s)$. The formal definition of this function is rather complicated and will be omitted here. For our purposes, it suffices to say that $\zeta(s)$ is a complex-valued function, which is very important in number theory.

Some of the solutions of equation

$$\zeta(s) = 0 \qquad\qquad (3.12)$$

are quite straightforward. It is known, for example, that equation (3.12) is satisfied for $s = -2, -4, -6, \ldots$ and all other *even* negative integers (such solutions are commonly referred to as *trivial*). In contrast, the complete set of *nontrivial* solutions is by no means obvious, and its identification has been the subject of intense research for over a century. Riemann's conjecture was that all such solutions have real parts equal to 1/2, but this has never been rigorously proved.

An obvious empirical argument in favor of Riemann's hypothesis is the fact that it was found to be true for as many as 10^{13} solutions of equation (3.12).[9] However, since there are actually infinitely many such solutions, no finite number of tests can guarantee that this conclusion holds in general. This observation is not a mere technicality. To see why that is so, we should note that certain properties of the zeta function are often described in terms of functions such as $\log(\log x)$, which grow very slowly as x increases. This means, for example, that in order to obtain $\log(\log x) = 4$, x would have to take the enormously large value $x = 10^{10,000}$, which exceeds the processing capability of standard mathematical software packages (Matlab cannot represent such a number, and classifies it as infinitely large). With that in mind, it is conceivable that the first nontrivial solution of equation (3.12) whose real part is *not* 1/2 could have an extremely large imaginary part - so large, in fact, that it would not be detectable by any known numerical algorithm.

In the absence of a theoretical proof, some mathematicians have resorted to a kind of reasoning that is often used in experimental science. The idea proposed by Good and Churchhouse [10] in the 1960s provides a typical illustration of this approach. To better understand the logic behind their proof, consider an arbitrary integer n and its decomposition into prime factors (such a factorization always exists, although it can be difficult to find in practice). Based on this decomposition, we can define a function $\mu(n)$ as

$$\mu(n) = \begin{cases} 0, & \text{if some of the prime factors are } repeated. \\ 1, & \text{if there is an } even \text{ number of distinct prime factors.} \\ -1, & \text{if there is an } odd \text{ number of distinct prime factors.} \end{cases}$$

According to this definition, a number like $n = 30$ obviously corresponds to $\mu(30) = -1$, since it can be factorized as $2 \cdot 3 \cdot 5$. Proceeding in a similar manner, it is easily verified that numbers such as $n = 15$ and $n = 20$ produce $\mu(15) = 1$ and $\mu(20) = 0$, respectively (by virtue of the fact that $15 = 3 \cdot 5$ and $20 = 2 \cdot 2 \cdot 5$).

It has long been known that the Riemann hypothesis is equivalent to the statement that the sum

$$M(N) = \sum_{n=1}^{N} \mu(n) \tag{3.13}$$

increases in a particular way as $N \to \infty$.[11] With that in mind, Good and Churchhouse considered a slightly different sum

$$M^*(N) = \sum_{n \in \Omega} \mu(n) \tag{3.14}$$

where Ω is a set of N *randomly* chosen integers (instead of the orderly set $\Omega_0 = \{1, 2, \ldots, N\}$ that was used in (3.13). If Ω is chosen in this way, it can be shown that $M^*(N)$ satisfies the desired condition with a probability equal to one, which is remarkably similar to the result that we would like to establish.

It can be argued, of course, that similarity of form and numerical simulation are valid verification criteria in science and engineering, but not in mathematics where rigorous proof remains the norm. Nevertheless, this line of reasoning can provide a legitimate basis for arguing that the Riemann hypothesis is very *likely* to be true. To see why this is so, suppose we were to tabulate the function $\mu(n)$ for a sufficiently large number of integers. A careful analysis of the obtained values would show that $\mu(n)$ exhibits no regularity at all. In view of the apparent randomness of this function, it is perfectly reasonable to assume that there is no substantial difference between choosing sets Ω and Ω_0 in the summation of $\mu(n)$.[12] As a result, what holds for $M^*(N)$ most probably holds for $M(N)$ as well.

Although formalists tend to reject arguments of this type as heuristic and unacceptable, many practicing mathematicians find them quite compelling. Human experience (including science) suggests that it would be highly unusual if Riemann's hypothesis turned out to be incorrect, given the existing experimental evidence and the striking similarity between $M^*(N)$ and $M(N)$. This example strongly suggests that the actual practice of mathematics involves much more than rigorous proofs and infallible logic. There is clearly room for intuition, experience and even a certain degree of uncertainty, just as in all other areas of human inquiry.

> "If mathematics describes an objective world just like physics, there is no reason why inductive methods should not be applied in mathematics just the same as in physics." *Kurt Gödel*[13]

The similarities between modern mathematics and the natural sciences are even more apparent in the domain of computer-aided experimentation. In dealing with physical systems that are too complex for direct measurements or observations, scientists routinely postulate mathematical models and use numerical simulation techniques to predict the dynamic behavior. Such experiments often point to previously unobserved phenomena, and suggest entirely new research directions.

In recent years, this "experimental" approach has gained considerable support among mathematicians as well. Computer simulation is now widely recognized as a tool that can point to new and unexpected results. Of course, such results still require a rigorous proof before they can be accepted as more than conjectures. Nevertheless, they often provide a very accurate idea of what it is that we should attempt to prove.

> "Experiment has always been, and increasingly is, an important method of mathematical discovery. ... The early sharing of insights increases the possibility that they will lead to theorems: an interesting conjecture is often formulated by a researcher who lacks the

technique to formalize a proof while those who have the techniques at their fingertips have been looking elsewhere." [14]

In making this point, Borwein and Bailey [15] give the example of one of their undergraduate students who computed the first 500,000 terms of the sum

$$\sigma = \sum_{k=1}^{\infty} \left(1 + \frac{1}{2} + \ldots + \frac{1}{k} \right) \cdot k^{-2} \tag{3.15}$$

In analyzing the data, he observed that $\sigma \approx 17\zeta(4)/4$, where $\zeta(s)$ represents Riemann's zeta function (see Example 3.3). This experimental discovery ultimately led to a whole class of analytical formulas, which connect sums of the form

$$\sigma(m, n) = \sum_{k=1}^{\infty} \left(1 + \frac{1}{2} + \ldots + \frac{1}{k} \right)^m \cdot k^{-n} \tag{3.16}$$

to different values of the zeta function.

The following example further illustrates the capacity of experimental mathematics to anticipate new and unexpected theoretical results.

Example 3.4. Consider the simple first-order difference equation

$$x(k + 1) = x^2(k) + p \tag{3.17}$$

where $x(0) = 0$ and p is a complex-valued parameter. In the following, we will be interested in identifying all the values of p for which the sequence $\{x(k)\}$ remains *bounded*. This problem was first studied by mathematician Benoit Mandelbrot, who proposed a simple method for estimating such a set. His algorithm was based on the fact that sequence (3.17) is guaranteed to diverge if the condition $|x(k)| > 2$ holds for some finite value of k. Making use of this property, Mandelbrot defined an upper limit k_{max} for the number of iterations, and plotted all the values of p for which

$$|x(k)| \leq 2 \tag{3.18}$$

for all $k \leq k_{max}$. The result of this procedure was an extraordinarily elaborate figure in the complex plain, which is known as the *Mandelbrot set*.

The form of this set becomes increasingly fascinating as we "zoom in" and examine its structure in greater detail. One of the many intricate patterns that emerge under these circumstances is shown in Fig. 3.4. It is safe to say that such an unusual and complicated mathematical object could never have been discovered without the help of computer simulation.

"The Mandelbrot set is the most complex object in mathematics, its admirers like to say. An eternity would not be enough time to

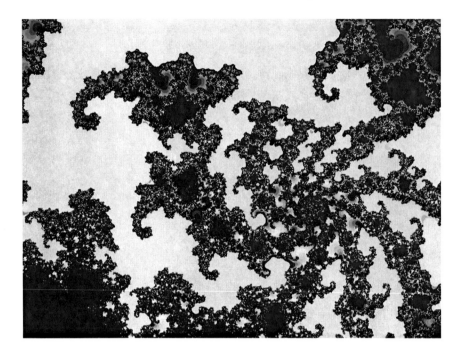

Fig. 3.4. A segment of the Mandelbrot set.

see it all, its disks studded with prickly thorns, its spirals and fila-
ments curling outward and around ... Examined in color through the
adjustable window of a computer screen, the Mandelbrot set seems
more fractal than fractals, so rich is its complication across scales."
James Gleick [16]

Another surprising property of the Mandelbrot set was observed by mathematician David Boll in 1991. Boll studied equation (3.17) with parameter values
of the form

$$p = -0.75 + j\varepsilon \qquad (3.19)$$

(where ε is a small number) and established that the sequence $\{x(k)\}$ diverges
for *any* $\varepsilon > 0$. Although this was not an unexpected result, he also found that
the number of iterations needed to reach $|x(k)| > 2$ varies with ε in a peculiar
way. This dependence is shown in Table 3.2, from which it is easily verified that
the product $\varepsilon N(\varepsilon)$ approximates π with an error less than $\pm\varepsilon$.

Why π would show up in this context remains a mystery. There is nothing in
equation (3.17) to suggest this property, nor have there been any other theoretical explanations to date. What is more important, however, is the recognition
that this is one of many instances where computer experiments uncovered a
surprising result, and defined a new mathematical problem. Without this kind
of preliminary investigation, we would probably have no idea that chaos and

ε	Number of iterations, $N(\varepsilon)$	$\varepsilon N(\varepsilon)$
0.1	33	3.3
0.01	315	3.15
0.001	3143	3.143
$\vdots$	$\vdots$	$\vdots$
0.0000001	31415928	3.1415928
$\vdots$	$\vdots$	$\vdots$

Table 3.2. The correlation between $\varepsilon N(\varepsilon)$ and π.

fractals even exist. The same holds true for many other complex phenomena that are now the subject of rigorous theoretical research.

Results like the ones described above have had a significant impact on the philosophy of mathematics. Doron Zeilberger, for example, speculates that in the not so distant future mathematical papers may have a very different form from the one that we are accustomed to. He illustrates this idea with the following hypothetical abstract:

> "We show in a certain precise sense that the Goldbach conjecture is true with probability larger than 0.99999 and that its complete truth could be determined with a budget of 10 billion." [17]

If this ever comes to pass, it will become difficult to separate the methods of mathematics and the natural sciences. Perhaps this has always been the case in practice, although one could hardly tell that by simply reading mathematical publications. Philosopher Imre Lakatos repeatedly made this point in his classic book *Proofs and Refutations*:

> "The Euclidean methodology has developed a certain obligatory style of presentation. I shall refer to this as the "deductivist style"... In the deductivist style mathematics is presented as an ever-increasing set of eternal, immutable truths. The deductivist style hides the struggle, hides the adventure. The whole story vanishes, the successive tentative formulations of the theorem in the course of the proof-procedure are doomed to oblivion while the end result is exalted into sacred infallibility. [18]

3.3 Notes

1. Douglas Hofstadter, *Gödel, Escher, Bach: An Eternal Golden Braid*, Basic Books, 1999.

2. Bertrand Russell, *Mysticism and Logic and Other Essays*, Rowman and Littlefield, 1981.

3. This system was considered in Hofstadter, *Gödel, Escher, Bach: An Eternal Golden Braid*.

4. This particular excerpt was taken from Philip Davis and Reuben Hersh, *The Mathematical Experience*, Birkhauser, 1981.

5. Kurt Gödel, *On Formally Undecidable Propositions of Principia Mathematica and Related Systems*, Dover Publications, 1992.

6. Bertrand Russell, *Portraits from Memory: And Other Essays*, Textbook Publishers, 2003.

7. Freeman Dyson, *Infinite in All Directions*, Harper and Row, 1988.

8. Jonathan Borwein and David Bailey, *Mathematics by Experiment*, A. K. Peters, 2004.

9. The current benchmark for the number of computed solutions was set by Gourdon and Demichel in 2004.

10. I. J. Good and R. F. Churchhouse, "The Riemann hypothesis and pseudo-random features of the Möbius sequence," *Mathematics of Computation*, **22**, pp. 857–861, 1968.

11. The specific condition for $M(N)$ is that

$$\lim_{N\to\infty} \left[\frac{M(N)}{N^{(1/2+\varepsilon)}} \right] < \infty \qquad (3.20)$$

for *any* $\varepsilon > 0$. Note that this is *not* equivalent to saying that $M(N)$ grows like $N^{1/2}$ when $N \to \infty$. For example, if $M(N)$ were to have the form $M(N) = N^{1/2} \log N$, (3.20) would still hold since

$$\lim_{N\to\infty} \frac{\log N}{N^\varepsilon} = 0 \qquad \forall \varepsilon > 0 \qquad (3.21)$$

The same holds true for a variety of other functions, such as $N^{1/2} \log(\log N)$, $N^{1/2} \log(\log(\log N)))$, *etc.*

12. Because of the apparently random character of $\mu(n)$, there is no clear preference between different ways of choosing set Ω. In that sense, Ω_0 is simply one of many possible choices, with no special properties.

13. Quoted in: Borwein and Bailey, *Mathematics by Experiment*.

14. D. Epstein, S. Levy and R. de la Llave, "About this journal," *Experimental Mathematics*, **1**, 1992.

15. Borwein and Bailey, *Mathematics by Experiment*.

16. James Gleick, *Chaos: Making a New Science*, Penguin Books, 1988.

17. D. Zeilberger, "Theorems for a price: Tomorrow's semi-rigorous mathematical culture," *Notices of the American Mathematical Society*, **40**, pp. 978-981, 1993.

18. Imre Lakatos, *Proofs and Refutations: The Logic of Mathematical Discovery*, Cambridge University Press, 1977.

Chapter 4

Quantum Mechanics

"Anyone who has not been shocked by quantum physics has not understood it." *Niels Bohr* [1]

4.1 The State of a Quantum Particle

Much of what goes on in the mysterious world of quantum mechanics is difficult to describe in the language of classical physics. Concepts such as position, velocity or energy, for example, are still applicable, but must be used with great care in order to avoid erroneous interpretations. Although our mathematical models provide a great deal of information about the way microscopic particles behave, we must concede that this knowledge is often best articulated in terms of analogies and metaphorical descriptions:

> "Quantum theory provides us with a striking illustration of the fact that we can fully understand a connection though we can only speak of it in images and parables." *Werner Heisenberg* [2]

Most textbooks on quantum mechanics begin by introducing the notion of the "state" of a particle, so it would be logical for us to do the same. Perhaps the best way to describe what this term means is to consider an analogy with a coin toss. We know from experience that however many times such an experiment is performed, the coin will always be observed in one of two states: heads-up or tails-up. These are the only two outcomes that our instruments can record, if we exclude the extremely unlikely scenario in which the coin lands on its edge. A practically minded individual would probably accept such a conclusion without further discussion, and would consider the matter settled at this point. Someone more philosophically inclined, however, might be tempted to ask the following question: What "state" was the particle in while it was still in the air? This is obviously not the kind of information that our hypothetical measurement devices can provide. And yet, it is perfectly reasonable to assume that the coin must be in *some* physical state before it lands.

If we allow for the existence of such a state, we must also concede that it is fundamentally different from the other two that we can measure. It is a state in which all options remain open, and the coin behaves as if it has not yet "made up its mind" regarding what face it will reveal to us. To the average reader, this may perhaps sound more like a Zen koan than like physics. Nevertheless, there is a clear analogy between our coin and a quantum particle. Like the coin, in certain situations the particle will have a *definite* (and measurable) value of some physical quantity (such as position or momentum, for example). In others, it will *not*.

To get a better sense for what this really means, we first need to recognize that the "state" of a quantum particle is generally described in terms of a *wave function*. This complex-valued function is usually denoted by the Greek letter ψ, and its temporal evolution is governed by Schrödinger's equation. Given a set of initial conditions, this equation allows us to compute ψ at any point x and at any time t. The problem, however, is that the wave function provides only partial information about the particle. If, for example, we happen to be interested in measuring its momentum, all that we can deduce from Schrödinger's equation are the *probabilities* of different outcomes. We have no way of knowing what value will actually be recorded. This is not simply a reflection of our ignorance and inability to measure physical quantities at the microscopic level - it is an *irreducible uncertainty*, which is built into the very foundations of quantum mechanics. If you find all this somewhat confusing and counterintuitive, you are in excellent company - Einstein had difficulty accepting this interpretation as well, and repeatedly suggested that its proponents were making too much out of an incomplete physical theory. He was ultimately proven wrong, but the resolution of this matter had to wait until the early 1980s.

It might be helpful at this point to consider a simple example, which can clarify the connection between the wave function and actual measurements. To that effect, let us assume that we are interested in some physical quantity Q which has only *two* observable values, q_1 and q_2 (much like our hypothetical coin). In the following, we will represent the states that correspond to these two values by wave functions ψ_1 and ψ_2, respectively. According to the rules of quantum mechanics, a measurement of Q will produce value q_1 whenever the system is in state ψ_1, and q_2 when it is in state ψ_2. But what can we say about a state ψ in which Q does *not* have a definite value? Such a state obviously differs from both ψ_1 and ψ_2, and corresponds to a situation where a measurement could produce *either* q_1 *or* q_2. It is as if the particle is still "making up its mind," just like the coin while it was still in the air.

The fact that a particle is in a state of "indefinite" Q does not prevent us from speculating on how it might behave when exposed to our measurement devices. The wave function ψ actually contains some useful information, which allows us to assign probabilities to different outcomes. To see how this information can be extracted, we should first note that ψ can always be represented as a *linear combination* of states ψ_1 and ψ_2. This linear combination has the form

$$\psi = \alpha_1 \psi_1 + \alpha_2 \psi_2 \tag{4.1}$$

where α_1 and α_2 are complex coefficients. Because of this property, a particle described by wave function ψ is said to be in a *state of superposition*.

It is interesting to note in this context that equation (4.1) is similar to the sort of expressions we use to represent vectors in a plane. Indeed, it is well known that any two dimensional vector can always be decomposed as

$$\vec{V} = v_1 \vec{e}_1 + v_2 \vec{e}_2 \tag{4.2}$$

where $\vec{e}_1$ and $\vec{e}_2$ are *unit vectors* along the x and y axes, while v_1 and v_2 correspond to the *projections* of $\vec{V}$. Such a situation is schematically illustrated in Fig. 4.1.

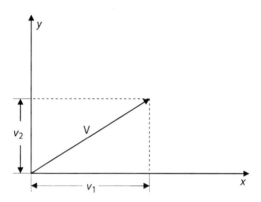

Fig. 4.1. Vector $\vec{V}$ and its projections.

The coefficients α_1 and α_2 in equation (4.1) can be viewed as "projections" in their own right, which have a specific physical interpretation. In quantum mechanics, the quantities $|\alpha_1|^2$ and $|\alpha_2|^2$ are understood as the *probabilities* that q_1 and q_2 will be recorded when we perform a measurement. Since q_1 and q_2 are the only possible outcomes in this case, the probabilities must also satisfy

$$|\alpha_1|^2 + |\alpha_2|^2 = 1 \tag{4.3}$$

The above discussion naturally leads us to one of the most controversial issues in quantum mechanics - the so-called "measurement problem." A key concept in this debate is the "collapse" of the wave function, which is a term used to describe the transition from a state of superposition to a state in which quantity Q has a definite value. It is by no means clear how (or why) an act of observation causes this change to occur. Some have suggested that it may be a result of human interference, which is inevitable in the process of measurement. Niels Bohr, for example, believed that the measurement device is a purely "classical" entity, while the particle itself conforms to the laws of quantum mechanics. From that perspective, the collapse of the wave function could be viewed as a sort of "bridge," which somehow enables the particle to cross the gap that separates the world of quantum mechanics from the classical domain.

A rather different interpretation is the so-called "many worlds" view of quantum mechanics, which was proposed by physicist Hugh Everett in the 1950s. This theory questions the notion of a "collapse," and claims that *each* possible state actually materializes in a parallel universe. At the time of the observation the observer's mind mysteriously "splits" and becomes simultaneously present in *multiple universes* (which have no contact with each other). Thus, in any given world we can perceive exactly one state of the particle, but have no conscious awareness that this experience takes place on multiple fronts.

Physicist Wojciech Zurek[3] of Los Alamos Laboratories recently proposed yet another theory, which is based on the idea of 'decoherence'. To understand what this means, we should recall that only *coherent* light exhibits wave properties such as interference. It is well known that a coherent beam of light is hard to produce – this usually requires a laser, or some similar device. Furthermore, coherence is a very 'fragile' property, since it can be destroyed by any natural disturbance. Zurek has argued that a similar situation arises in quantum mechanics. When particles exhibit something analogous to 'coherence', their quantum wave properties (such as superposition of states) are apparent. However, when they 'decohere' (due to environmental noise, for example), the quantum system turns into a classical one, with a single identifiable state that can be measured. This, of course, is a hypothetical conclusion, but many physicists see it as a plausible explanation for the collapse of the wave function.

4.2   The Uncertainty Principle

Heisenberg's Uncertainty Principle is one of the best known results of quantum mechanics. In simple terms, this principle maintains that we cannot simultaneously detect the position and momentum of a quantum particle without an error. The problem that we are dealing with here is not unlike measuring the distance between two cities. If cities were points, this distance could be determined with arbitrary precision. However, since they actually have finite dimensions, it is impossible to reduce the measurement error below a certain threshold.

In order to gain a better understanding of the Uncertainty Principle, we must first consider how expression (4.1) changes if quantity Q is assumed to have more than two measurable values. If the set of possible outcomes consists of *discrete* values $\{q_1, q_2, \ldots, q_n\}$, the decomposition of the wave function retains pretty much the same form, the only difference being that there are now n terms in the expansion. This holds true even if there are infinitely many possibilities, in which case ψ can be expressed as

$$\psi = \sum_{i=1}^{\infty} \alpha_i \psi_i \tag{4.4}$$

The situation becomes considerably more complicated when Q has a *continuous* set of measurable values. An example of such a quantity is the position of a particle, which can take on *any* real value if there are no physical constraints. For our present purposes, it suffices to observe that in such cases we need to deal with *probability distributions* instead of discrete probabilities. An example of such a distribution is shown in Fig. 4.2, in which $p(q)$ denotes the so-called "probability density" associated with value q.

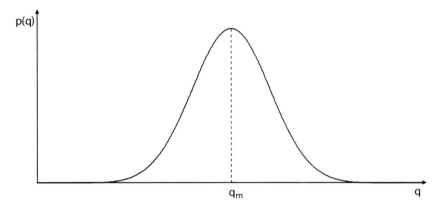

Fig. 4.2. Example of a probability distribution.

What does this graph really tell us? If we wanted to compute the probability that Q will take a value in the interval $[a, b]$ (that is, $a \le q \le b$), we could do so by calculating the shaded area shown in Fig. 4.3. Note that the total area under the curve must always equal 1, since it represents the probability that $-\infty \le q \le \infty$.

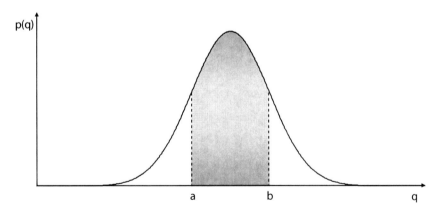

Fig. 4.3. Probability that q lies in the interval $[a\ b]$.

Distributions like the one shown above are characterized by two quantities:

1) The value q_m, which corresponds to the maximum of function $p(q)$.

2) The "spread" (or *dispersion*) of the curve, which we will denote in the following by Δq.

Since dispersions of physical variables play a particularly important role in Heisenberg's Uncertainty Principle, we will need to consider them in greater detail. We begin by contrasting two radically different situations, which are illustrated in Figs. 4.4 and 4.5, respectively. In the first case, it is reasonable to say that the particle is very close to having a definite value of Q, since all possible outcomes are tightly concentrated around q_m. In the second case, however, we can say nothing of the sort. Fig. 4.5 indicates that there is, in fact, a very broad range of different values for q, with q_m being only slightly more likely than the others.

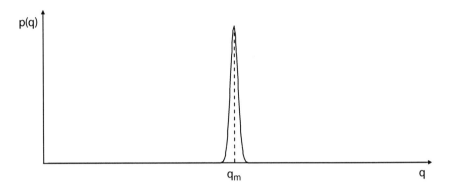

Fig. 4.4. Distribution with a small spread.

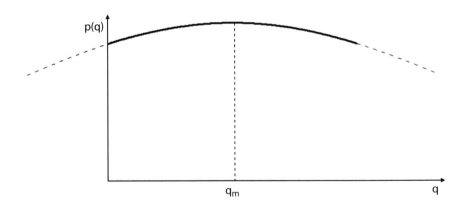

Fig. 4.5. Distribution with a large spread.

The most widely known version of Heisenberg's principle concerns the position and momentum of a particle (denoted in the following by x and p, respectively). Both of these quantities have a continuous spectrum of possible values. If the particle happens to be in a state in which *neither x nor p* have a definite value, the probability distributions of these variables would typically look something like the one shown in Fig. 4.2, with dispersions Δp and Δx, respectively. What Heisenberg recognized was that these two dispersions are *not* independent, and must satisfy the inequality

$$\Delta x \Delta p \geq \frac{\hbar}{2} \tag{4.5}$$

where $\hbar$ is Planck's constant divided by 2π.

How should such a result be understood? To see that, let us imagine a particle that is enclosed in a tiny box which allows for a very small Δx. Such a particle would obviously have a well defined (albeit not entirely definite) location, but its momentum would be almost completely uncertain. Indeed, given inequality (4.5), Δp would have to be very large to compensate for the small dispersion Δx. If we were now to measure the momentum of such a particle, it is quite possible that the recorded value would be very large (as suggested by the graph in Fig. 4.5). It is as if the particle becomes "claustrophobic" in response to its spatial confinement, "bouncing around" violently within the limits of the box.

A common interpretation of inequality (4.5) is that we cannot simultaneously measure the position and velocity of a particle. While such a statement is technically correct, it fails to capture the essence of Heisenberg's result. What the uncertainty principle really tells us is that the more we know about the position of a particle, the less we can say about its momentum (and vice versa). In the extreme case when Δx is small and Δp is very large, one could legitimately argue that the concept of momentum (which is a classical attribute) can not be meaningfully associated with the particle. This is the point where the language of Newtonian physics becomes inadequate (and can even be thoroughly misleading).

It is important to recognize that Heisenberg's principle is not limited to position and momentum, and can be extended to other observable physical quantities as well. An important example is the expression

$$\Delta E \Delta t \geq \frac{\hbar}{2} \tag{4.6}$$

where ΔE represents the possible change in energy, and Δt is the time over which this change takes place. Inequality (4.6) suggests that over very short periods of time energy can fluctuate dramatically, even to the point of producing a short lived particle-antiparticle pair in empty space (according to special relativity, pure energy can be converted into matter and vice versa). One of the consequences of this fact is that "vacuum" is actually teeming with activity, and should be thought of as empty only *on the average*. In Chapter 6, we shall see that the existence of this microscopic "quantum frenzy" has far reaching implications, and is one of the main obstacles for the development of a consistent theory of quantum gravity.

4.3 The Dual Nature of Matter

From its very inception, modern science has operated under the assumption that there exists an objective "external" world which can be studied using reliable theoretical and experimental techniques. Quantum mechanics undermines this hypothesis, and suggests that the physical attributes of a particle actually depend *on the way we choose to observe it.* We could decide, for example, to perform an experiment which would prove that an electron has wavelike characteristics (such as interference and spatial distribution). Alternatively, we could set up our experimental apparatus in a way that would clearly demonstrate the electron's corpuscular nature. Although these two results appear to be at odds with each other, modern physics claims that they are both legitimate, and represent *equally valid* manifestations of the electron's "physical reality." An electron, in other words, exhibits *particle – wave duality*, and the "face" that it will reveal to us depends on our method of investigation.

Would it then be correct to say that different measurement strategies sometimes lead to contradictory conclusions about the fundamental properties of matter? Quantum mechanics certainly allows for such a possibility. And since the choice of measurement strategy is a *conscious* human decision, we must also acknowledge the prospect that there may be no "real" physical world "out there" whose existence is completely independent of the observer. However bizarre it may seem, it is entirely plausible to assume that the observer participates in creating the reality that is observed, since the questions that we ask ultimately determine what we will see.

This apparent paradox prompted Niels Bohr to formulate his famous Principle of Complementarity, which states that the existence of incompatible measurement results needn't necessarily constitute a logical contradiction. Instead, it can be viewed as an indication that reality has multiple aspects, which sometimes seem to be in conflict with each other, but are ultimately unified on a level that eludes conventional reasoning. Bohr articulated his thoughts on this subject in a way that sounds almost mystical:

> "The opposite of a correct statement is an incorrect one, but the opposite of a profound truth is another profound truth." *Niels Bohr* [4]

The trail of events that lead to the discovery of particle-wave duality can be traced back to the beginning of the nineteenth century. In 1801, English physicist Thomas Young performed a groundbreaking experiment that definitively established the wave-like nature of light. Young showed that beams of coherent light [5] emanating from two narrow slits produce an interference pattern on the screen, which is a property that is unique to waves. The success of this experiment appeared to settle a long-standing dispute on the nature of light, dating back to the 17th century. In these early debates, Isaac Newton maintained that light consisted of particles, while his rival Christian Huygens argued in favor of waves.

The triumph of Huygens' view turned out to be relatively short-lived. By the early 1900s, Newton's outlook was fully vindicated by the discovery of the photoelectric effect, which proved that light also exhibits corpuscular properties. This lead to the inevitable conclusion that light has a *dual* nature, whose different aspects emerge under different circumstances.

Inspired by this result, French physicist Louis de Broglie put forth the bold conjecture that *all* matter possesses a dual nature. Although his predictions were thoroughly counterintuitive, they were soon verified in practice by American scientists Davisson and Germer, who studied the scattering of electrons in nickel. They considered an equivalent of Young's experiment in which the incoming beam was composed of *electrons* rather than light, and observed the same kind of interference pattern on a phosphorescent screen.

Although Davisson and Germer provided us with indisputable empirical evidence of particle-wave duality, some aspects of this strange property are still not fully understood. One of the most important unresolved questions in this context concerns the *nature* of the waves that are associated with particles. In the case of light, these waves can obviously be attributed to oscillations of the electromagnetic field. But what kind of waves could possibly be associated with an electron? The question remains undecided to this day. To illustrate why this is so, in the following we will present two very different (and seemingly contradictory) explanations. In comparing their merits, we should keep in mind that they are *equally valid* from a scientific perspective, since both produce accurate predictions.

Probability Waves

The first explanation dates back to the 1920s, and was proposed by German physicist Max Born. We know that according to Schrödinger's equation the probability that a particle (say an electron) will be found at a certain point x is determined by the wave function $\psi(x, t)$. From a mathematical standpoint, this function represents a legitimate wave that propagates through space. With that in mind, Born assumed that probability waves behave similarly to light, and exhibit *interference properties.*

Born's hypothesis implies that the shape of the probability wave in the double slit experiment is sensitive to *both* slits, although the electron itself passes through only one of them. This means that the waves will add up at certain points on the screen, increasing the probability of the electron being registered there. In other points the waves will subtract, and the probability will be reduced. If the experiment is repeated with many individual electrons, the resulting probability distribution implies that certain points on the screen will be hit often, while others will be hit much less frequently. The combined effect will produce an interference pattern that is consistent with the one obtained by Young.

Feynman's View

In the 1960s, Richard Feynman proposed a model that challenged the assumption that an electron must pass either through one slit or the other. Instead, he suggested that in moving between two points each electron actually traverses *every* possible trajectory simultaneously. He then showed that the combined average over these paths produces the same probabilities as the ones calculated using the wave function.

On first glance, this seems like a very strange conclusion. Indeed, how can a particle possibly follow multiple paths at the same time? One would expect that such a hypothesis could easily be refuted by recording whether the electron actually passed through slit A or through slit B. This, however, is easier said than done. Any attempt to perform such a measurement would inevitably disturb the system, and would alter the outcome of the experiment (among other things, the interference pattern would disappear). It follows, then, that for all practical purposes we *cannot* identify a specific trajectory for the particle, and are therefore in no position to experimentally disprove Feynman's claim.

Another apparent difficulty with this explanation comes from our actual experience, in which all macroscopic objects follow a single, well defined path. This appears to be clearly inconsistent with Feynman's conjecture. It turns out, however, that this is not the case, since for larger objects the combined contributions of all paths *except one* cancel out. As a result, the proposed approach does predict a *single* trajectory in our everyday world.

The fact that these two theories are considered to be equivalent despite their fundamental differences may seem paradoxical, but such situations are not uncommon in science. For the most part, they arise when the phenomenon under investigation eludes our conventional categories and forms of explanation. In that respect science is not unlike theology, which often describes the same transcendent reality in different (and sometimes mutually exclusive) ways. We will make use of this similarity in Section 10.4, where we examine the problem of religious diversity.

4.4   The EPR Paradox

The view that quantum mechanics is inherently indeterministic is commonly known as the Copenhagen interpretation (in honor of one of its greatest advocates, Danish physicist Niels Bohr). Although this outlook quickly gained support among scientists, it was never without its fair share of critics. Einstein, who was one of its most vociferous opponents, voiced his dislike for this interpretation with the famous phrase: "God does not play dice with the universe." He rejected Heisenberg's view that a particle in a state of superposition is just a collection of possibilities, and suggested instead that the microscopic world conforms to a set of *deterministic* laws, whose precise nature has yet to be determined. In the 1930s, Einstein and his two graduate students, Podolski and Rosen, produced what they thought was a decisive argument against Bohr's interpretation of quantum theory.[6] This argument, which became known as the EPR paradox, was the subject of heated debate for nearly fifty years. When it was finally resolved in the early 1980s, Einstein was proven wrong.

In order to explain the nature of the EPR effect, we first need to describe what is meant by the "spin" of a particle. This is a strange property which arises when relativistic effects are incorporated into quantum mechanics. In order to gain an intuitive understanding of this concept, imagine a ball with a unit radius that spins around its own axis. All points that belong to such a ball obviously experience circular motion (except, of course, those that are *on* the axis itself). Given the angular velocity of this object, classical physics allows us to uniquely determine the corresponding angular momentum.

The problem with extending this concept to quantum mechanics stems from the fact that particles cannot be viewed as microscopic "balls." Instead, they are formally represented as *point-like* entities which have no physical dimensions. Such objects clearly cannot rotate around their own axis in any conventional sense, and it is hard to see how one could associate an angular momentum with them. And yet, that is precisely what quantum mechanics does when it talks about the "spin" of a particle.

In dealing with quantum spin, it is important to keep in mind that this property has no counterpart in the world of our everyday experience. We can perhaps interpret it as something that is mathematically similar to conventional rotation, but is ultimately different in kind. The nature of this difference is clearly illustrated by the fact that spin is an *intrinsic* characteristic of the particle, just like its mass or charge. This is quite unlike classical physics, where the angular momentum explicitly depends on the velocity of the object.

Spin can be characterized by two pieces of information:

1. *The amount of spin*, which is fixed for a given particle. Particles whose spin is an odd multiple of $\hbar/2$ are referred to as *fermions*, while those whose spin is an integer multiple of $\hbar$ (including zero) are know as *bosons*.

2. *The direction of the spin*, which can be clockwise or counter clockwise. These two possibilities correspond to 'spin up' and 'spin down' states, respectively.

To illustrate how quantum spin is treated mathematically, let us consider an isolated electron, which has the smallest possible amount of spin for a fermion (that is, $\hbar/2$). What we don't know about the electron's spin is its *direction*, and this is something that we can measure. Since there are only two possible outcomes, the states of *definite spin* can be described by wave functions ψ_1 and ψ_2. If the electron is *not* in one of these two states, its wave function must be expressed as their superposition

$$\psi = \alpha_1\psi_1 + \alpha_2\psi_2 \qquad (4.7)$$

In this "state of superposition," $|\alpha_1|^2$ and $|\alpha_2|^2$ represent the probabilities of measuring 'spin up' and 'spin down', respectively.

It is interesting to note that when spin measurements are performed, we normally pick an *arbitrary* axis. We have no reason whatsoever to expect that this guess will coincide with the actual axis around which the particle is spinning. However, due to the collapse of the wave function, the particle will automatically align with the axis that we have chosen, and $|\alpha_1|^2$ and $|\alpha_2|^2$ provide the probabilities of observing 'spin up' and 'spin down' states.

Now that we have a basic understanding of quantum spin, we can proceed to examine what the EPR paradox is all about. It is perhaps best described in terms of an experiment in which an atom with spin zero disintegrates into two smaller particles (say A and B) that fly off in different directions. According to quantum mechanics, prior to a measurement the state of each particle ought to be a superposition of "up" and "down" spins. Note, however, that if we happen to measure spin "up" for particle A, it is guaranteed that we will subsequently measure spin "down" for B, regardless of the distance between them. This is due to the conservation of angular momentum, which holds both in classical and quantum physics (keep in mind that in this case the spin of the "parent" atom was zero).

The EPR paradox relates to the fact that measuring the spin of particle A somehow places particle B in a state of definite spin as well, although B itself has not yet been "disturbed" (B could actually be very far from A when the measurement occurs). There is no physical mechanism that would allow information about measurements made on particle A to be "instantaneously transmitted" to particle B, given that no signal can propagate faster than light. Based on this apparent contradiction, Einstein argued that quantum mechanics is, in fact, an incomplete theory. He suggested that particles A and B were really subject to "hidden" deterministic laws, and are both in a state of *definite spin* which is decided at their "birth." He further hypothesized that we cannot predict exactly what this state will be, but can assign a "classical" probability to each outcome.

Although Einstein and his collaborators were not aware of this at the time, there is actually a way to explicitly test the validity of their conjecture. This test is based on certain properties of classical probabilities which were discovered by physicist John Bell in 1964. To get a sense for what Bell's result means, suppose that particles A and B have *definite* spins, and consider all possible combinations in the three spatial directions. In Table 4.1, these states are summarized as triplets (s_x, s_y, s_z), which are necessarily complementary for the two particles along each axis, due to the conservation of momentum.

	Particle A	Particle B
1	$(x \uparrow, y \uparrow, z \uparrow)$	$(x \downarrow, y \downarrow, z \downarrow)$
2	$(x \uparrow, y \uparrow, z \downarrow)$	$(x \downarrow, y \downarrow, z \uparrow)$
3	$(x \uparrow, y \downarrow, z \uparrow)$	$(x \downarrow, y \uparrow, z \downarrow)$
4	$(x \uparrow, y \downarrow, z \downarrow)$	$(x \downarrow, y \uparrow, z \uparrow)$
5	$(x \downarrow, y \uparrow, z \uparrow)$	$(x \uparrow, y \downarrow, z \downarrow)$
6	$(x \downarrow, y \uparrow, z \downarrow)$	$(x \uparrow, y \downarrow, z \uparrow)$
7	$(x \downarrow, y \downarrow, z \uparrow)$	$(x \uparrow, y \uparrow, z \downarrow)$
8	$(x \downarrow, y \downarrow, z \downarrow)$	$(x \uparrow, y \uparrow, z \uparrow)$

Table 4.1. Possible spin combinations.

As an illustration, let us consider the probability that particle A has spin 'up' in the x-direction (*i.e.* $x \uparrow$) and that B has spin 'up' in the y-direction (*i.e.* $y \uparrow$). Out of the eight possible scenarios in the table, only combinations 3 and 4 correspond to this particular outcome. Consequently, the overall probability

$P(x\uparrow, y\uparrow)$ can be expressed as

$$P(x\uparrow, y\uparrow) = \frac{n_3 + n_4}{N} \tag{4.8}$$

where n_3 and n_4 represent the number of generated particles with combinations 3 and 4, respectively, while N denotes the total number of particles. Following the same procedure, probabilities $P(x\uparrow, z\uparrow)$ and $P(z\uparrow, y\uparrow)$ can be computed as

$$P(x\uparrow, z\uparrow) = \frac{n_2 + n_4}{N} \tag{4.9}$$

and

$$P(z\uparrow, y\uparrow) = \frac{n_3 + n_7}{N} \tag{4.10}$$

Observing that

$$\frac{n_3 + n_4}{N} < \frac{n_2 + n_4}{N} + \frac{n_3 + n_7}{N} \tag{4.11}$$

we now have that

$$P(x\uparrow, y\uparrow) < P(x\uparrow, z\uparrow) + P(z\uparrow, y\uparrow) \tag{4.12}$$

which is one of several possible formulations of *Bell's inequality*.

This simple result proved to be crucial for verifying the validity of "hidden variable" theories, since it provided a condition that the particles must satisfy if their behavior is truly classical. The definitive experiment along these lines was conducted by Alain Aspect and his collaborators in 1981.[7] Their measurements conclusively established that Bell's inequality was *violated* in this case, which confirmed that quantum phenomena do *not* lend themselves to strictly deterministic descriptions.

Given such a result, the only consistent explanation of the EPR paradox is that no matter how far apart A and B are, their common "parentage" prevents us from treating them as separate entities. In technical terms, this means that the two particles shouldn't be described by separate wave functions, but rather by a *single* one (even if they end up in different galaxies). When particle A is observed, the *joint* wave function collapses, causing particle B to assume a definite spin value as well. According to this interpretation quantum mechanics must be viewed as a *non-local* theory, and all microscopic phenomena are somehow connected into a non-decomposable whole.

> "As a result of Bell's theorem and the experiments it stimulated, a supposedly purely philosophical question has now been answered in the laboratory: There *is* a universal connectedness. ... Any objects that have ever interacted continue to instantaneously influence each

other. Events at the edge of the galaxy influence what happens at
the edge of your garden." *Bruce Rosenblum* [8]

Some Concluding Thoughts

Now that we have developed a basic understanding of the strange and counterintuitive laws that govern the microscopic world, what general conclusions
can we draw? One of the most striking lessons that quantum mechanics has to
offer is that exclusive opposites can coexist in nature (provided, of course, that
we broaden our notion of "existence"). We have also seen that physical reality
contains an irreducible element of uncertainty, and that the universe appears to
be interconnected in ways that defy the human imagination. Mathematics can
perhaps provide us with a language to describe these phenomena, but it cannot
really "explain" them in any meaningful way. Niles Bohr was absolutely right
- quantum physics *is* shocking, even to those who have studied it for years. In
that respect, it resembles certain mystical practices, which force us to abandon
traditional concepts and invite us to explore a world that cannot be fully grasped
by the intellect.

4.5   Notes

1. http://en.wikiquote.org, 2005.

2. Werner Heisenberg, *Physics and Beyond: Encounters and Conversations*,
   Harper & Row, 1971.

3. W. H. Zurek, "Decoherence and the transition from quantum to classical",
   Physics Today, **44**, pp. 36-44, October 1991.

4. Quoted in: Heisenberg, *Physics and Beyond: Encounters and Conversations.*

5. When we say that light is 'coherent', we mean that its rays must be in
   phase.

6. A. Einstein, B. Podolsky and N. Rosen, "Can quantum-mechanical description of physical reality be considered complete?" *Physical Review*,
   47, pp. 777-780, 1935.

7. A. Aspect, P. Grangier and G. Roger, "Experimental tests of realistic local
   theories via Bell's theorem," *Physical Review Letters*, **47**, pp. 460-463,
   1981.

8. Bruce Rosenblum and Fred Kuttner, *Quantum Enigma*, Oxford University
   Press, 2006.

Chapter 5

The Theory of Relativity

> "What purpose does it serve for me that time and space are exactly the same thing? I ask a guy what time it is, and he tells me six miles?" *Woody Allen* [1]

In the first two decades of the twentieth century, Einstein produced not one but *two* remarkable theories that changed science in a profound way. The first of these, known as *special relativity*, examined how physical processes might appear to observers who are in relative motion. Einstein's response to this question brought about a radical shift in our understanding of space, time, matter and energy. The second theory (which goes by the name of *general relativity*) provided a fundamentally new explanation for the effects of gravity, and opened the door for a systematic exploration of the universe. In the following, we will provide a brief overview of both theories, and consider one of their crowning achievements - the discovery of "black holes."

5.1 Special Relativity

Galileo Galilei was the first scientist to mathematically connect observations made in reference systems that are in relative motion. His results are quite straightforward, and provide a good starting point for describing Einstein's more abstract ideas. The simplest way to introduce the so-called "Galilean transformations" is to consider a moving object which is viewed from two different coordinate systems, A and B (as shown in Fig. 5.1). We will assume that one of these systems is stationary, while the other moves with a *constant* velocity v_R in the direction of the x-axis.

If $D(t)$ represents the relative distance between the origins of the two coordinate systems, it is obvious from Fig. 5.1 that

$$x_A(t) = x_B(t) + D(t) \qquad (5.1)$$

where $x_A(t)$ and $x_B(t)$ denote the position of the object as seen by the two observers at time t. Since $D(t) = v_R t$, the relationship between $x_A(t)$ and $x_B(t)$

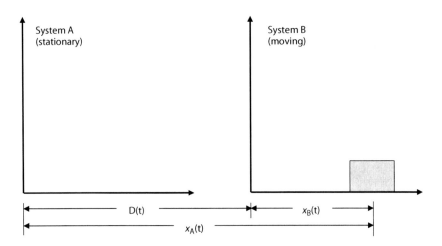

Fig. 5.1. Two coordinate systems in relative motion.

becomes

$$x_A(t) = x_B(t) + v_R t \tag{5.2}$$

In interpreting equation (5.2), it is important to recognize that in the Galilean model time is *identical* in all coordinate systems. This assumption allows us to directly relate the velocities and accelerations that are recorded by our two observers as

$$v_A(t) = v_B(t) + v_R \tag{5.3}$$

and

$$a_A(t) = a_B(t), \tag{5.4}$$

respectively. From expressions (5.3) and (5.4) we can conclude that the velocity of the object will generally be *different* for the two observers, but the acceleration remains the same (as do all forces).

When analyzing coordinate transformations such as the one described above, it is usually of interest to identify quantities that remain *unchanged* by them. In the case of Galilean transformations, it is easily verified that the *distance* between two objects exhibits this property. Indeed, if two coordinate systems move with a relative velocity v_R, the distance between objects P and Q can be expressed as

$$x_A^{(p)} - x_A^{(q)} = x_B^{(p)} + v_R t - \left[x_B^{(q)} + v_R t \right] = x_B^{(p)} - x_B^{(q)} \tag{5.5}$$

This sort of invariance allows us to introduce a common measure of distance for *all* frames of reference (such a measure is referred to as a *metric*).

The Galilean approach was simple and worked extremely well in the context of Newtonian mechanics. In the late 1800s, however, it was observed that Maxwell's equations (which describe the propagation of light) are *not* invariant to such transformations. This posed a conceptual difficulty, since basic electromagnetic phenomena ought to be unaffected by the relative motion of the observers. In 1905, Einstein proposed a solution to this problem which lead to a fundamental change in the way we perceive space and time. His theory (which is known as *special relativity*) grew out of two remarkably simple postulates:

Postulate 1. The laws of physics must be identical for all observers moving at constant velocities.

Postulate 2. The speed of light is the same for all observers moving at constant velocities.

While the first of these postulates seems perfectly reasonable, there is something profoundly counterintuitive about the second one. Our entire experience suggests that objects moving toward us appear *faster*, while those moving away seem to be *slower*. Why would light be the only exception? There was no obvious reason for making such an assumption, apart from Einstein's intuition. Nevertheless, his conjecture turned out to be correct, and ultimately lead to some remarkable discoveries.

In order for Einstein's two postulates to hold, it was necessary to introduce a new coordinate transformation under which Maxwell's equations would remain unchanged. By the early 1900s, it was established that this requirement is satisfied by the so-called *Lorentz transformation*, which is based on the assumption that two observers will record *different times* and *different spatial locations* for the same event. This is quite unlike the Galilean paradigm, which assumed that all observers record the *same* time for an event (the difference being only in the spatial coordinates).

The Lorentz transformation allows us to connect coordinates (t_A, x_A) and (t_B, x_B) in a rather straightforward manner. What is interesting in this case is that position x_A recorded by observer A generally depends on *both* t_B and x_B (as does time t_A). The fact that temporal and spatial coordinates somehow become "entangled" as we move from one reference system to another has far-reaching consequences. Among other things, it suggests that space and time cannot be treated as separate entities, and should instead be viewed as components of a unified four dimensional *spacetime*.

Although the Lorentz transformation is just a mathematical operation, it turns out that it has some very real physical manifestations. One of the best known examples of this kind is the so-called *Lorentz contraction*, which implies that a stationary observer will record a *shorter* length for an object than his moving counterpart. This effect is negligible at velocities that are much smaller than the speed of light, and is therefore not something that we can normally experience. However, if we could move at the speed of light relative to the object, the distinction between its "front" and "back" would completely vanish.

Unfortunately (or perhaps fortunately), there is no way for us to do that, for reasons that will be explained shortly.

An even more striking illustration of the way in which Lorentz transformations alter our conventional ideas about time and space is a phenomenon known as *time dilation*. To get a sense for what this means, imagine two observers, one stationary, and the other travelling in a space ship. We will assume that both observers record the time when the ship departs, and when it arrives at its destination. What will their clocks show? According to the Lorentz transformation, the elapsed time recorded by the moving observer is *shorter* than the time measured by the stationary one. In the extreme case when the spaceship approaches the speed of light, time would effectively stop for the moving observer. For him, the past, present and future would merge into what can be described as an "eternal present moment" (a notion that is often encountered in accounts of mystical experiences).

It is important to recognize at this point that the impact of Lorentz transformations is by no means limited to space and time. It turns out, for example, that this relatively simple mathematical result also has a profound effect on our understanding of matter and energy. To see why this is so, let us first recall that the Lorentz transformation was originally introduced to ensure the invariance of Maxwell's equations under coordinate changes. However, according to Einstein's postulate, *all* laws of physics must have this property, including those of Newtonian mechanics. It was soon recognized that such a requirement poses a significant problem, since Newton's laws (in their classical formulation) are *not* invariant under Lorentz transformations. Einstein proposed to resolve this difficulty by developing a new "relativistic" mechanics, whose formalism reflects the idea that spacetime is a unified four dimensional entity.

In making the appropriate adjustments, Einstein took special care to ensure that his relativistic equations are consistent with Newton's laws at low velocities. This constraint reflected the fact that classical mechanics was a well tested and highly successful theory under "normal" conditions. In view of that, it is fair to say that Einstein developed a model that *includes* and *transcends* Newton's description of the physical world. The two theories become significantly different only for objects that move at velocities comparable to the speed of light.

One of the most famous results of Einstein's relativistic mechanics is the formula

$$E = mc^2 \tag{5.6}$$

which relates the total energy of a particle to its mass (c being the speed of light). In interpreting this expression, it is important to keep in mind that mass is *not* constant, and depends on the particle's velocity as

$$m = \frac{m_0}{\sqrt{1 - v^2/c^2}} \tag{5.7}$$

It is true, of course, that this dependence is barely noticeable in the world of macroscopic phenomena (where the condition $v \ll c$ is usually satisfied). However, it becomes very prominent in describing the motion of subatomic particles, whose velocities often approach the speed of light. The extreme case when $v = c$ is particularly interesting in this context, since it appears to lead to a contradiction. Indeed, if we allow $m_0 \neq 0$ and $v = c$ to hold simultaneously, this would obviously result in an infinitely large mass (which is physically impossible). Einstein recognized that the only way to avoid this difficulty was to assume that particles with non-zero mass *cannot* move at the speed of light.

Another important implication of equations (5.6) and (5.7) is that a particle has a certain amount of energy even when it is not moving (*i.e.* when $v = 0$). This is quite unlike classical mechanics, where the energy of a free particle at rest is assumed to be zero. The fact that this *intrinsic* potential energy (whose value is $E_0 = m_0 c^2$) is directly proportional to the mass of the particle allows us to view matter and energy as equivalent and interchangeable quantities.

To understand the full ramifications of equation (5.6), we should recall that classical physics has two *separate* conservation laws for mass and energy. The first of these laws implies that energy in a closed system can change its form (*e.g.* from kinetic to electrical or from nuclear to thermal), but the total amount must remain *unchanged*. The law of conservation of mass, on the other hand, claims that although the chemical makeup of a substance could undergo a drastic transformation (as in the case of burning), the initial amount of matter must stay the same.

Einstein's brilliant insight was that these two laws *cannot* be separated, and that matter can be converted into energy (and vice versa). Since c^2 turns out to be a very large number, equation (5.6) suggests that the sudden conversion of a small amount of matter can create an enormous amount of energy. This equation also implies that under certain conditions pure energy can be converted into matter. This phenomenon is well known in quantum mechanics, which predicts that elementary particles can appear literally out of 'nowhere,' simply as a result of energy fluctuations (see inequality (4.6) and the related discussion).

5.2 General Relativity

According to Newton's theory of gravity, any two objects will attract each other with a force that is proportional to the product of their masses. If we denote these masses by m_1 and m_2, respectively, the corresponding gravitational force can be expressed as

$$F = \gamma \frac{m_1 m_2}{r^2} \tag{5.8}$$

where γ is the so-called *gravitational constant*, and r represents the distance between the two centers of mass. To keep things as simple as possible, in the following we will focus only on objects whose center of mass coincides with their geometric center (this is the case, for example, with homogeneous spherical bodies).

The model described above was widely accepted for more than two centuries, although it was unable to explain how gravitational influences are instantaneously transmitted over large distances. Newton himself believed that his equations adequately described the mechanism of the interaction, but not its causes:

> "It is inconceivable that inanimate brute matter should, without the mediation of something else, which is not material, operate upon and affect other matter without mutual contact. ... Gravity must be caused by an agent acting constantly according to certain laws; but whether this agent be material or immaterial, I have left to the consideration of my readers." *Isaac Newton* [2]

Einstein's new theory of gravity (which became known as *general relativity*) represented a radical departure from Newton's classical paradigm. One of its most original contributions was the assumption that space is an *active* entity, rather than a mere "container" into which material objects are placed. Einstein's conjecture was that spacetime and matter actually *interact*, and that this process can be described in precise mathematical terms. According to this hypothesis, spacetime becomes curved in the presence of matter (or energy), and its geometric form is determined by the way in which the masses are distributed. In such a framework, any change in the distribution of matter would produce a change of curvature, but *not* instantaneously. The propagation of gravitational disturbances was assumed to proceed at the speed of light, with a necessary transient period until spacetime settles into its new configuration.

What exactly do we mean when we say that "spacetime is curved"? A natural way to approach this question would be to recognize that we happen to be living on the surface of a sphere, which is definitely not flat. In such a space, *curvilinear* coordinates such as angles are often a more natural way to describe the location of an object than their Cartesian counterparts. The notion of curved time is somewhat more exotic, and requires a bit of imagination. The simplest interpretation of this phenomenon is that time can flow faster or slower, depending on the local geometric properties. It turns out, for example, that in the neighborhood of a massive object (where spacetime is significantly curved) time "slows down". As we move away from such an object, the curvature becomes less pronounced and time "accelerates". Due to this effect, a clock positioned in a remote satellite would run faster than clocks on earth.

An analogy that is often employed to describe Einstein's notion of gravity asks us to imagine empty space as a flat, elastic surface. If we were to place a heavy ball onto it, this surface would obviously experience a significant deformation, and our flat two dimensional space would become "curved." If we were now to roll a marble in the direction of the ball, its trajectory would obviously be affected by the deformation. Exactly what would happen to the marble in this case would depend on its initial position and velocity. Most probably, it would simply "swerve" as it encounters the "dent" that surrounds the ball. However, under certain special circumstances, it might actually enter into an orbit similar to that of planets in our solar system.

There are several "lessons" that can be drawn from this analogy. Perhaps the most important one is that the motion of objects in space is not the result of a "gravitational force," as Newton had postulated. Instead, it is a consequence of the curving of space in the neighborhood of massive objects. According to this interpretation, spacetime should be viewed as an "active" medium which interacts with matter. The nature of this interaction is perhaps best summarized in John Wheeler's famous line:

"Spacetime grips mass, telling it how to move;

And mass grips spacetime, telling it how to curve." [3]

As is the case with all theories that challenge our basic concepts, general relativity was initially met with considerable skepticism. To many it seemed like an imaginative and elegant mathematical exercise which had little to do with physical reality. It turned out, however, that this theory had a number of testable predictions, which were soon verified. The first of these concerned certain anomalies in the orbit of planet Mercury, which had puzzled astronomers for many years. Unlike all other planets in our solar system, Mercury's trajectory is not a stationary ellipse, but rather a "rotating" one (such an orbit is often referred to as non-reentrant). Prior to the emergence of general relativity, this phenomenon was attributed to the presence of a yet-to-be discovered planet, which was even given a name (it was called Vulcan). Einstein's theory provided a very different explanation, and was able to accurately predict the observed trajectory.

The definitive experimental confirmation of general relativity came in 1919. In that year, British astrophysicist Arthur Eddington photographed a number of stars during a total solar eclipse (which made them briefly visible during the day). The positions that he recorded were different from the "usual" ones that these stars occupied in the night sky. This discrepancy was precisely what general relativity predicted, since beams of light originating from the stars are expected to "bend" as they pass by the "dent" in spacetime caused by the sun. The angle of this deflection can be calculated from Einstein's equations, and the obtained values matched perfectly with Eddington's observations. It is important to emphasize in this context that Newton's theory of gravity cannot account for such a phenomenon. To see why this is so, it suffices to consider equation (5.8), which implies that the classical "force" of gravity ought to affect only entities that have nonzero mass. This is clearly *not* the case with light, since its carrier particles (photons) are massless.

One of the most remarkable features of general relativity is the fact that it provides a mathematical framework for studying the universe as a whole. It goes without saying, of course, that such a framework is necessarily incomplete, since there is no way to accurately describe a system of such complexity. Nevertheless, Einstein's equations allow us to formulate a meaningful approximate model of the universe, and to make certain predictions. In evaluating such a model, we ought to be aware of some key assumptions that are built into it. Perhaps the most important one is the so-called *Cosmological Principle*, according to which the universe is assumed to be *homogeneous* and *isotropic* at any given point in time. These two rather technical terms imply that the large-scale structure of the universe is assumed to be uniform, and that what we see through our telescopes should not depend on where we are, or which direction we are looking in.

The assumptions contained in the Cosmological Principle give rise to an exact solution of Einstein's equations which is known as the *Robertson-Walker metric*. This solution allows for three possible characterizations of the universe:

1) The universe could be *closed*, in the sense that it has a positive curvature. In that case, it would be destined to collapse into a single point in a "Big Crunch" (a process that is the exact opposite of the Big Bang).

2) The universe could be *critical*, with a zero curvature. A "flat" universe like this is highly improbable, since such a state could be disturbed by even the most minute inhomogeneity.

3) The universe could be *open*, with a negative curvature. Such a universe would be destined to expand limitlessly.

Current astronomical observations suggest that the universe is very close to the critical divide - so close, in fact, that we cannot determine with any certainty whether it is open or closed. The scales are so delicately balanced that small variations in matter density throughout the universe make all the difference. Unfortunately, we cannot hope to ever know the exact nature of these variations, since some of the necessary information is contained in parts of the universe that are (and will remain) invisible to us. This places a strict limit on what science can say about this question.

What we *do* know, however, is that the universe is currently expanding. This conclusion follows from numerous astronomical observations, which clearly indicate that galaxies are moving away from each other. To see what this really means, imagine for a moment that the universe is like the surface of a balloon, and that galaxies resemble coins which are "glued" to it. In this analogy, humans would play the role of tiny two dimensional creatures that move along the surface of the balloon, but have no physical contact with its interior and exterior. If air is now let into the balloon, the surface will obviously expand and the coins will move apart, much like real galaxies do.

It is interesting to note that this analogy suggests a universe of finite size, but with no real "end." As in the case of the earth's surface, there is no boundary point from which we could "fall off," nor does it make sense to look for a specific point where the universe "began." It does make sense, however, to ask *when* it began. Indeed, if the universe is truly expanding in time (as all our data suggests), we could hypothetically run Einstein's equations "in reverse" until we reach an instant when it shrinks to a single point. In cosmology, this point when space and time effectively cease to exist is known as a *singularity*. It is from such a singularity that our universe presumably emerged, in an explosion of unimaginable proportions known as the Big Bang.

General relativity cannot explain why a singularity would explode, or why it would result in this particular universe and not some other. Under such extreme conditions, its mathematical formalism simply breaks down. Most scientists are inclined to believe that singularities do not really exist in nature, and that the need for this concept will be eliminated once Einstein's theory is successfully combined with quantum mechanics. This does not mean, however, that the Big Bang model should be dismissed as inaccurate. Despite its apparent limitations, this theory provides a viable description of how the universe evolved after the first 10^{-43} seconds (when quantum effects played a prominent role). It also makes a number of testable predictions, the most important of which is the existence of left-over radiation from the initial explosion. This "background" radiation was detected by Arno Penzias and Robert Wilson in 1964, and its spectrum was found to be in perfect accordance with theoretical expectations. Their discovery, together with several other experimental results, gave considerable credibility to the Big Bang model, which is now a widely accepted theory.

In weighing the merits and weaknesses of the Big Bang model, we should also note that the initial singularity is by no means the only obstacle that it has faced. One of its earliest problems had to do with the assumption that the universe is homogeneous and isotropic (which greatly simplifies the mathematics associated with Einstein's equations, and allows for exact solutions). Over the years, many scientists questioned whether such an assumption is actually physically justified. Indeed, although the Cosmological Principle is generally consistent with our astronomical observations, we must bear in mind that our telescopes can see only as far as the *visible horizon* (which is the region of the universe from which radiation has had time to reach us since the Big Bang). Whatever lies beyond remains a mystery.

In addition to being a practical inconvenience, the existence of the horizon poses a serious theoretical problem. Standard Big Bang theory explained the observed uniformity of the universe by the fact that it emerged from a miniscule "nugget" where all particles were very close together. As such, they could interact and homogenize. The difficulty with this argument is that it does not allow enough time for the homogenization to take place. In the 1980s, Alan Guth [4] suggested a way to resolve this long standing problem. He managed to find a solution of Einstein's equations that allows for a brief period of exponential expansion, which is called *inflation*. According to inflation theory, prior to this accelerated expansion the universe was only *locally* uniform. In other words, what is now our observable universe may have originated from a tiny segment of the primordial "nugget", which was small enough for particles to interact and homogenize. The other segments (with possibly very different properties) also expanded, but are now located beyond our visible horizon. Among other things, this hypothesis accounts for the amazing regularity that we see in all directions of space, while allowing for the possibility that the invisible universe is very different. These other regions could easily have laws and constants of nature that are nothing like the ones that we observe.

It is important to recognize in this context that although our observable universe is indeed extremely "smooth," it is not *completely* uniform. If the distribution of matter happened to be perfectly symmetric, it could never have clustered into stars and galaxies, and life as we know it would not have emerged. Scientists currently believe that the small variations in density and temperature that we observe today originate from quantum fluctuations that occurred in the immediate aftermath of the Big Bang (see Section 4.2). Although these fluctuations are extraordinarily small, it is entirely possible that they could have been "amplified" in the course of inflationary expansion.

In recent years, Guth's inflationary model has given rise to a number of new hypotheses regarding the origin of the universe. One such possibility was proposed by Andrei Linde[5], who suggested that inflation tends to "reproduce", with each expanding nugget literally repeating the process that brought it into being. Since there are small differences within each nugget, the "parent patches" needn't be perfectly homogeneous and can produce slight variations in the fundamental physical constants. Lee Smolin[6] took this line of reasoning a step further, by proposing an "evolutionary" theory of the universe. He observed that both the Big Bang and the center of a black hole are characterized by an enormous (theoretically infinite) density of matter. Based on that, he suggested that every black hole is like a "seed" for a new universe, which emerges through a Big Bang-like explosion. This, of course, is something that we could never verify, due to the shielding effect of the black hole's "event horizon" (in the following section, we will describe what this term means).

Smolin further suggests that the characteristics of the new universes depend on the "parent" black hole, which in turn depends on the features of the star that formed it. The "baby" universe would exhibit similar physical constants, with slight variations. In this process, universes which are better suited for the formation of black holes would reproduce more prolifically, ultimately leading to progeny that is optimized for black hole production.

5.3   Black Holes

One of the most fascinating and counterintuitive predictions of general relativity is the existence of black holes. The discovery of these strange objects is a prime example of the power of human imagination, which can anticipate realities that are far removed from our senses.

> "I do not know of any other example in science where such a great extrapolation was successfully made solely on the basis of thought."
> *Steven Hawking*[7]

To get a proper sense for what black holes really are, we must first examine certain physical processes that take place in stars. We begin by observing that the interior of a star is a lot like a nuclear reactor, in which hydrogen atoms are fused into helium. The energy released by this process creates an outward pressure, which is strong enough to balance the gravitational attraction among the gas particles. As long as these conditions hold, the star remains in a stable state. At some point, however, the nuclear "fuel" in its core will become depleted, and gravity will begin to gain the upper hand. This is when the star begins to *collapse.*

Exactly how this process will end depends on a number of factors. If the star is very massive, its center might collapse while its outside shell of gas explodes, scattering large amounts of matter across the universe. Such a massive explosion is known as a *supernova*. It is entirely possible, on the other hand, that a star will

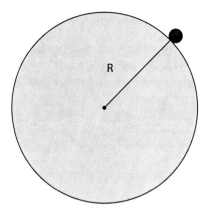

Fig. 5.2. A star with an object on its surface.

retain its original mass as it "shrinks." In such cases, there are two plausible scenarios - either the star will stop contracting at some point (and become a white dwarf or a neutron star), or it will continue to collapse until its entire mass is compressed to the size of a point. The second of these two possibilities corresponds to a *black hole*.

Whether or not a star will become a black hole ultimately depends on its mass. According to current estimates, this ought to occur when the original amount of matter in the star exceeds $3M_S$ (M_S stands for *solar mass*, which is a standard unit in astrophysics). It is important to keep in mind, of course, that this is just an estimate, since precise equations for extremely dense states of matter have yet to be formulated. Nevertheless, the above value gives us a pretty good idea of conditions that must be in place for the creation of these unusual celestial objects.

One of the principal characteristics of black holes is their immense gravitational "pull." In a certain region around the center of the hole, this pull is so strong that *nothing* can escape, including light (hence then name "black hole"). As a result, such objects are completely invisible to us. In order to explain this phenomenon, it is helpful to temporarily return to Newton's classical theory of gravity, despite the fact that it is not the appropriate model for studying black holes. According to this theory, the force of gravity between two spherical objects of masses m and M is

$$F = \gamma \frac{mM}{r^2} \tag{5.9}$$

where r represents the distance between their centers. A special case arises when M is the mass of a star, and m corresponds to a small object on its surface (as illustrated in Fig. 5.2). In such cases, r is equal to the radius of the star, which we will denote in the following by R.

For our purposes, it would be interesting to see what would happen if the star were to shrink to half of its original size, while retaining its entire mass

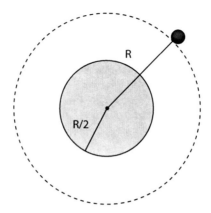

Fig. 5.3. The situation after the star shrinks.

(which is similar to what happens in the process that gives rise to a black hole). In that case, we would have a situation like the one shown in Fig. 5.3, with $r = R/2$. It is important to recognize that the force felt by the smaller object would remain *unchanged*, since it is still at a distance R from the center of the star. However, if the object were somehow attached to the surface, it would feel a force

$$F = \gamma \frac{mM}{(R/2)^2} = 4\gamma \frac{mM}{R^2} \tag{5.10}$$

which is four times larger than what it was before. This means (among other things) that it would be much more difficult for such an object to "escape" from the surface of the star. The velocity with which it would have to be fired in order not to be pulled back to the surface is known as the *escape velocity*, which obviously becomes larger as the star's radius decreases.

When the radius of the star reduces to some critical value R_C, the force of gravity on its surface becomes so large that the object will not be able to escape even if it is launched at the speed of light. As the mass of the star continues to be compressed, R_C becomes the boundary of a imaginary sphere from which nothing can escape any more, even light itself. This boundary defines what is known as the *event horizon*, whose formation marks the "birth" of a black hole.

We should keep in mind, of course, that this is only a classical analogy for what is otherwise an extremely complex physical phenomenon. The proper mathematical framework for any "real" explanation would have to be Einstein's theory of general relativity. Nevertheless, there is enough commonality between the two models to allow for certain meaningful conclusions. Newton's classical theory suggests, for example, that the gravitational "force" at the event horizon ought to remain unaffected by a further compression of the star's mass. General relativity says pretty much the same thing, except for the fact that it uses the term "spacetime curvature" instead of "force." Both models imply that a hypothetical observer hovering near the event horizon would experience the same gravitational effects that he felt at the point when the black hole was formed. If R_C happens to be sufficiently large, this effect could actually be fairly moderate.

In a somewhat more morbid version of this thought experiment, we could imagine an observer who is "attached" to the surface of the shrinking star. He would obviously experience a progressively stronger gravitational pull as the radius R approaches zero. Interestingly, it is not the gravitational force per se that would have the most dramatic impact on this unfortunate individual - it is the *change* in this force from one point to another. To see this more clearly, consider two points on the observer's body whose radial distance from the center of the star is R_0 and $R_0 - \Delta R$, respectively. The difference in the gravitational force that he would feel at these two points is

$$\Delta F = 2\gamma \frac{mM}{R^3} \Delta R \qquad (5.11)$$

In practical terms, equation (5.11) implies that the forces pulling at his head and feet would become dramatically different as R approaches zero. This phenomenon, which is known as the *tidal effect*, would ultimately tear the observer apart.

Moving our discussion in a less depressing direction, we should now consider the fate of the star itself. According to general relativity, it will continue to shrink until its entire mass becomes compressed into an unimaginably small volume. What happens then? Einstein's equations predict that the curvature of spacetime will become infinite, which is why it is referred to as a *singularity*. The intriguing question, of course, is whether such a thing can actually exist in nature. Scientists tend to get suspicious when their theories predict infinitely large quantities, and black holes are no exception to this rule. It is therefore not surprising that most physicists believe that a singularity is probably a point where general relativity breaks down, as does our conventional understanding of space and time. It is reasonable to assume that some future theory of quantum gravity might provide a better explanation of how matter behaves under such extreme conditions.

Evidence of Black Holes

Up to this point, our analysis focused mostly on theoretical speculations about black holes. But what can we say about actual evidence for their existence? Visual evidence is clearly out of the question in this case, since black holes emit no light whatsoever. With that in mind, we should first examine whether they have some properties that we might detect *indirectly*. It turns out that they do. In the following, we will consider several interesting phenomena related to black holes, each of which can provide us with important clues about their existence. To begin with, we should recall that these objects inevitably curve spacetime in their vicinity. One would therefore expect that waves of light passing close to them will experience some gravitational "bending" (as was the case in Eddington's experiment). This is clearly something that our measurement instruments can record.

Another measurable effect associated with collapsing stars is the so-called *gravitational redshift*. This phenomenon can be explained in part by the Doppler effect, which is associated with the propagation of waves. The usual framework for describing this effect involves a moving light source, whose emissions are recorded by a stationary observer. Although the frequency of the light produced by the source is always the same, what we perceive on the receiving end depends on whether it is approaching us or receding from us. In the former case, the observed frequency would be *higher* than its nominal value, and in the latter it would be *lower*.

How does all this relate to black holes? When a star collapses, its surface (from which light is emitted) contracts, and therefore effectively recedes from us. As a result, we should expect the recorded frequency to shift toward the lower end of the visible spectrum (which corresponds to red light). We should add in this context that the "redshift" is amplified by the gravitational pull of the collapsing star. As an illustration of this process, imagine photons that are trying to "escape" from its surface. As the star's radius shrinks, one would expect the photons to use more and more energy to overcome the increased gravitational pull. If they actually succeed in doing this, they will be left with a rather empty "gas tank." Since the energy of a photon is proportional to its frequency, we can conclude that the light emitted from the surface of the collapsing star will be shifted toward the red end of the spectrum.

When discussing potentially measurable characteristics of black holes, we should also mention one that has to do with their rotation. Not all black holes rotate, of course, but those that do can significantly affect particles of matter that are located in their neighborhood. To see how this phenomenon can provide us with additional clues for detecting black holes, imagine a solid sphere of mass M and radius R which is rotating around an axis through its center. For such an object we can compute the "classical" angular momentum as

$$L = I\omega \tag{5.12}$$

where ω is the *angular velocity*, and I is the *moment of inertia*. For a solid sphere like the one we are currently considering, it can be shown that

$$I = \frac{2}{5}MR^2 \tag{5.13}$$

If no external forces act on such a system, its angular momentum must be *conserved* (this is one of the fundamental laws of classical physics).

Let us now picture a rotating black hole as a spinning sphere of mass M. As this sphere contracts, its radius R decreases, as does its moment of inertia (according to equation (5.13)). If the angular momentum is to be preserved, this change must be compensated by a corresponding *increase* in the angular velocity ω. This means, in other words, that the star whose collapse created the black hole will rotate faster and faster as its dimensions become smaller.

How might a rapidly rotating black hole affect surrounding objects? To see this, imagine for a moment that spacetime resembles a "sticky" honey-like

substance. If a sphere rotates in such a medium, objects in its vicinity will begin
to spin alongside it. Obviously, the faster the rotation, the more pronounced
this effect will be. The area around a rotating black hole where surrounding
objects are affected by this sort of phenomenon is known as the *ergosphere*. If
we envision the event horizon as an imaginary sphere, then the ergosphere will
generally extend beyond it. The deviation from the event horizon is largest at
the equator, and smallest at the poles (where the deviation is actually zero).

Given all these theoretical predictions, have scientists actually discovered real
black holes in the universe? The consensus among astrophysicists is that they
have, although the evidence is necessarily circumstantial. To see what grounds
they have for believing this, we first need to consider where one might look
for good candidates. A very promising possibility are so-called *binary stars*,
which orbit as a pair around a common center of gravity (such systems are quite
common in our galaxy). More often than not, we can register both stars in the
pair, and can record their relative movements. It can happen, however, that one
of the objects in the pair is *invisible*, in which we have a good candidate for a
black hole. Note however, that such evidence is by no means conclusive, since
neutron stars (which are difficult to detect visually) can produce similar effects.

Another indication that the second object in a binary system might be a
black hole is the presence of intense x-ray radiation. Theoretical considerations
suggest that the source of this radiation could be particles of gas that the black
hole attracts from its binary partner. Before being absorbed, the gas spirals
toward the black hole, forming a structure known as an *accretion disk*. The
particles tend to heat up as they move toward the interior of the disc, and
radiate the accumulated energy in the form of x-rays.

The criteria outlined above have helped scientists identify a number of strong
candidates for black holes. Most such candidates fall into one of two categories:

1) *Stellar mass* black holes, whose mass is $3 - 10$ M_S.

2) *Supermassive* black holes, whose mass ranges from one million to several
billion M_S.

A very promising candidate for a *stellar mass* black hole is an x-ray source
known as Cygnus X-1. Cygnus X-1 is the invisible part of a binary system in
which the visible star is bright, but produces little radiation. The mass of this
object is estimated to be *at least* $3M_S$, which is the minimum requirement for
a black hole. Unfortunately, our current theories are not entirely precise about
this lower limit, so the evidence is still considered to be inconclusive. This is
true of several other candidates of similar mass, such as A0620-00 and XTE
J1650-500 (the names of these objects reflect their coordinates in the sky).

The case seems to be stronger for *supermassive* black holes, whose mass is
thought to be millions (and even billions) of times larger than that of the sun.
A particularly interesting possibility for detecting such objects involves *quasars*.
These "exotic" celestial objects are found near the limits of our observable universe, which means that the light coming from them was emitted soon after the
Big Bang. The unusual thing about quasars is that they are extraordinarily
bright, despite their relatively small size. If that were not the case, we most

probably wouldn't be able to detect them, given the enormous distance that separates us.

Scientists speculate that quasars could perhaps be accretion disks of enormously massive black holes. Such theories, however, are very difficult to verify directly. Much of what we currently believe about these objects is extrapolated from our observations of *active* galaxies, whose cores periodically undergo violent disturbances. A characteristic feature of this process are two narrow streams of gas that emanate from the center of the galaxy. There are now a number of theories that link such bursts to supermassive black holes.

Evidence from the Hubble telescope strongly suggests the presence of such a black hole in the center of galaxy M87 (in the constellation of Virgo). Although the mass of this object is estimated at about 3 billion M_S, it is by no means the largest one. In galaxy NJC 4061 there is actually a black hole candidate which is thought to be up to 9 billion times more massive than the sun. There may even be a supermassive black hole in our own galaxy. The most promising candidate is a powerful source of x-ray radiation known as Saggitarius A*, whose mass is estimated to be some 3 million M_S.

Concluding Thoughts

As a final point in our discussion of special and general relativity, it might be interesting to revisit the question that Woody Allen raises in the opening quote of this chapter. What difference does it make to you and me if space and time (or matter and energy) are essentially the same thing? There is no doubt that humans could happily go about their business without ever learning anything about this (which is precisely what most of us do). However, for those who are inclined to examine the rationality of religious teachings, it is helpful to know that our conventional perceptions of reality are often misleading, and do not come even close to some of the models that are currently being used in science. The notion that spacetime is somehow "curved," for example, or the fact that the difference between the past, present and future vanishes for observers moving at high speeds suggest that the physical world is structured in a way that defies "common sense." With that in mind, it would be unreasonable (and unfair) to ask theologians to produce easily understandable explanations for the transcendental phenomena that they explore. They, too, should be allowed to use concepts and ideas that have no counterpart in our everyday experience.

5.4   Notes

1. This line is from Woody Allen's film: *Anything Else*.

2. Quoted in: Brian Greene, *The Elegant Universe*, Vintage Books, 2000.

3. John Wheeler, *A Journey into Gravity and Spacetime*, W. H. Freeman and Co., 1990.

4. Alan Guth, *The Inflationary Universe*, Addison-Wesley, 1997.

5. A. Linde, "Particle physics and inflationary cosmology," *Physics Today*, **40**, pp. 61-68, September 1987.

6. Lee Smolin, *The Life of the Cosmos*, Oxford University Press, 1997.

7. Quoted in: Kitty Ferguson, *Prisons of Light*, Cambridge University Press, 1996.

Chapter 6

String Theory

"It used to be that as we were climbing the mountain of nature the experimentalists would lead the way. We lazy theorists would lag behind. Every once in a while they would kick down an experimental stone which would bounce off our heads. Eventually we would get the idea and we would follow the path that was broken by experimentalists. Once we joined our friends we would explain to them what the view was and how they got there. That was the old and easy way (at least for theorists) to climb the mountain. We all long for the return of those days. But now we theorists might have to take the lead. This is a much more lonely enterprise." *David Gross* [1]

Although quantum mechanics and general relativity are arguably the most successful scientific theories of the 20th century, it turns out that they cannot both be entirely correct. This inconsistency gives rise to one of the most profound conceptual problems of modern physics. Some have argued that the discrepancy is of little practical significance, since quantum mechanics focuses on microscopic phenomena while general relativity deals with massive objects and large distances. It is fair to say, however, that for many scientists such reasoning is clearly inadequate. This view is further supported by the fact that there are some situations where the two theories must actually be used *simultaneously*. A typical example is the center of a black hole, where a huge mass is crushed to the size of a point.

The joint application of quantum mechanics and general relativity often yields results that don't make physical sense, such as probability values that tend to infinity. To understand the root of the problem, we must recall Heisenberg's uncertainty principle as it pertains to energy and time

$$\Delta E \Delta t \geq \frac{\hbar}{2} \tag{6.1}$$

As noted in Chapter 4, this relationship allows for rather "wild" fluctuations of energy over very short periods of time. Since energy can be converted into matter (by virtue of $E = mc^2$), these fluctuations can often be large enough to produce particle - antiparticle pairs in empty space. In order for this to happen, the time interval Δt would have to be very small, which means that such pairs annihilate each other almost immediately. Nevertheless, their brief "appearance" in the material world has some significant consequences. It turns out, for example, that this "frenzy" of microscopic activity creates tiny distortions in spacetime, which are sometimes referred to as "quantum foam" (a term coined by physicist John Wheeler).[2] Despite the fact that these effects arise on scales smaller than the Planck length (which is 10^{-35} meters), their existence is significant because it violates the assumption that spacetime is "smooth", which is crucial for Einstein's formulation of general relativity. The problem is compounded by the fact that standard quantum mechanics treats particles as *points*, which means that they are small enough to be affected by such disturbances in the fabric of spacetime. This has been a key motivating factor for the development of string theory, which proposes a framework for overcoming the discrepancies between quantum mechanics and general relativity.

6.1   Some Basic Properties of Strings

According to standard particle physics, matter has a limited number of fundamental constituents. A typical example is an electron, which *cannot* be reduced to a collection of "simpler" ingredients. This, however, is not the case with protons and neutrons, which consist of three *quarks* that are combined in two different ways. Physicists have actually discovered six different types of quarks, as well as a number of other fundamental particles. As shown in Tables 6.1-6.3, these particles can be classified very neatly into three "families" based on their masses (the values in these tables are given as multiples of the mass of a proton).[3]

Interactions between fundamental matter particles can be modeled as exchanges of *virtual particles*, which typically remain in existence for very short periods of time. In quantum electrodynamics, for example, it is useful to envision electrons and protons as surrounded by "clouds" of photons, which can overlap when the particles are sufficiently close to each other. When that happens, the ensuing electromagnetic interaction can be described as a *mutual exchange of photons*. It turns out that this theoretical framework is quite general, and can

Particle	Mass
electron	5.4×10^{-4}
electron-neutrino	$< 10^{-8}$
up quark	4.7×10^{-3}
down quark	7.4×10^{-3}

Table 6.1. Family I.

Particle	Mass
muon	0.11
muon-neutrino	$< 3 \times 10^{-4}$
charm quark	1.6
strange quark	0.16

Table 6.2. Family II.

Particle	Mass
tau	1.9
tau-neutrino	$< 3.3 \times 10^{-2}$
top quark	189
bottom quark	5.2

Table 6.3. Family III.

be extended to include all four basic forces of nature. Thus, we can view strong nuclear interactions as exchanges of *gluons*, while weak nuclear and gravitational interactions are associated with exchanges of *weak gauge bosons* and *gravitons*, respectively.

The main contention of string theory is that *all* fundamental particles (both of the "material" and "virtual" variety) are composed of microscopic "strings" which can be loosely viewed as vibrating one dimensional rubber bands. Strings conceived in this way can vibrate with an unlimited number of resonant patterns, much like a guitar string of a fixed length. While the resonant patterns of the guitar string give rise to different musical notes, the vibrational patterns of microscopic strings correspond to different *particles*. This leads us to conclude that both matter and virtual particles have the *same* fundamental constituents, the only difference being the type of vibration. That is unification "par excellence".

The fact that strings have a finite size (which is on the order of Planck's length) suggests that they are not likely to be affected by the distortions of spacetime that are due to quantum foam. To get an intuitive understanding for why this is so, it suffices to observe that when two point-like particles interact gravitationally, the location of the interaction is a definite point in spacetime. As a result, the mathematical description of such an interaction will encounter all the problems associated with discontinuities of spacetime on scales below the Planck length. In contrast, when two strings interact, the location of the interaction is not precise (since they are extended objects). It turns out that this eliminates many of the underlying mathematical problems, most notably the infinite probabilities.

Strings and Supersymmetry

Early formulations of string theory permitted only vibrational patterns that correspond to particles with whole-numbered spins (that is, *bosons*). Among the possible vibrational patterns allowed by this model were some that produced *complex* masses. The corresponding particles, referred to as *tachyons*, posed a variety of conceptual difficulties that required fundamental modifications to the theory.

The tachyon problem was eventually eliminated by incorporating the notion of *supersymmetry*, which allowed for the inclusion of fermions as well as bosons. It is for this reason that the current theory uses the term *superstrings*. Unfortunately, it turned out supersymmetry can be included in not one but *five* different ways, each resulting in a different mathematical model. This non-uniqueness has been one of the main problems facing string theory since the mid 1980s.

Some recent results indicate that there may be an overarching framework that unifies the five theories. While a thorough discussion of this framework (which is known as *M-theory*) clearly exceed the scope of this book, it doesn't hurt to mention some of its main features. The starting point of M-theory is the observation that the equations of all five models are actually approximations, which are based on *perturbation theory*. The central idea behind the perturbation approach is to start with a good approximate model, and then iteratively refine it by including progressively smaller corrections. If the corrections are not small, however, this method collapses, and the resulting equations are not correct. M-theory exploits this possibility, as well as certain dualities between the five different versions of string theory. Although it is still incomplete, it introduces some very intriguing new possibilities. Perhaps the most striking implication is that strings can actually be p-dimensional (with $p > 1$). Such strings are rather whimsically referred to as *p-branes*. This generalization also allows for the possibility of unifying gravity with the other three basic forces at very high energies.

Nine Spatial Dimensions

Although the replacement of point-like particles by strings eliminated the problem of infinite probabilities, initial versions of string theory encountered a number of additional technical difficulties. One of the most prominent issues was the emergence of *negative* probabilities in certain calculations. In order to resolve this problem, theorists suggested a very bizarre remedy - according to their derivations, it was necessary to introduce *six additional spatial dimensions* into our three dimensional world. Although these dimensions are extremely small (on the order of the Planck length), they are large enough to allow the string to vibrate in nine independent spatial directions. It was established that under such circumstances, the mathematical problems associated with negative probabilities could be completely resolved.

This result is so strange and counterintuitive that it deserves a more detailed analysis. In order to do that, let us consider a one dimensional world which is inhabited by "point-like" creatures. One being could never overtake another in such a world, since they are all confined to a single line. Let us now imagine a minute modification, in which the line turns into an extremely thin cylinder. The surface of this cylinder now constitutes a two dimensional space, which gives rise to many new possibilities. Among other things, our "point-like" creatures could overtake each other by moving along the surface of the cylinder.

This simple thought experiment introduces a world that has one "regular" linear dimension, and a second (microscopic) "curled" one. Locating a particle in such a space requires two independent pieces of information - a *linear* coordinate x and a *curvilinear* coordinate θ. The idea of adding such microscopic curled dimensions to a regular "flat" space is of fundamental importance in string theory. The geometry of these added microscopic dimensions will obviously have a significant effect on the kinds of vibrations that are possible (which further determines the particle properties that are permissible in such a model).

The equations of string theory place certain constraints on the possible six dimensional forms. The constraints are not too severe - it turns out that there is an entire class of spaces called *Calabi-Yau* shapes that meet the requirements. [4] These shapes are virtually impossible to visualize, and are best described as very complicated six dimensional "knots". Nevertheless, their mathematical description is well defined. The main problem in this context is how to pick the "right" Calabi-Yau shape out of thousands of candidates. This choice is important, since the geometric properties of various Calabi-Yau shapes can be very different, and can lead to distinct vibrational possibilities. However, to date no reliable criteria have been developed for selecting the proper shape for the curled dimensions.

Despite the undeniable elegance of the underlying mathematics, for most of us it remains very difficult to imagine a world with nine spatial dimensions. Even if we accept such a counterintuitive notion, we cannot avoid wondering why only three of the dimensions are large enough to be experienced. This is one of the most perplexing questions facing string theory, and there is no consensus regarding the correct answer. One possible explanation for the differences between the dimensions is based on the cosmological theory of inflation. Supporters of this view have argued that the initial "nugget" of spacetime that emerged from the Big Bang actually contained nine spatial dimensions, but that only three of them expanded in the process of inflation. The other six remained curled up and microscopic, affecting only objects whose size is on the order of the Planck length or smaller. As such, these dimensions cannot be detected by our instruments, and have no direct bearing on the way we experience reality.

6.2   The Explanatory Power of String Theory

Perhaps the greatest appeal of string theory lies in its potential to theoretically reconcile quantum mechanics and general relativity. Beyond this fundamental feature, it also provides a number of intriguing insights into some of nature's most elusive phenomena. In the following we will focus on two such issues - the so-called "cosmic bounce" and the problem of black hole entropy. Readers who do not have a technical background should be aware that these topics are slightly more advanced, and may be difficult to grasp without some knowledge of physics. Such individuals might want to focus their attention on the philosophical questions that are raised at the end of this section.

The Cosmic Bounce

As noted in Chapter 5, cosmological models that follow from Einstein's general relativity predict that the universe will either expand forever or will collapse in a "Big Crunch". To get an idea of how string theory approaches this problem, let us once again consider a hypothetical universe in the form of an extremely thin cylinder. In such a universe, a string could move along the surface of the cylinder, much like a point-particle would. However, a string could also *wrap* itself around the cylinder (possibly multiple times), and slide along in the linear dimension. This is a kind of motion that has no counterpart in the world of point-particles.

A wrapped string has two different sources of energy, which are referred to as *vibrational* and *winding* energy. Vibrational energy is a result of the sliding motion. As the radius R of the cylinder *decreases*, the string is localized in an increasingly smaller space. By Heisenberg's principle, the dispersion of the string's momentum will then become *larger*, which increases the likelihood of higher energy values. As a result, the energy corresponding to this motion will have the form

$$E_1 = \frac{k}{R} \qquad k = 1, 2, \ldots \tag{6.2}$$

where k is known as the *vibration number* (the fact that k takes integer values reflects the quantization of energy).

The winding energy depends on the total length of the string, since this determines its mass. Obviously, the more windings there are, the higher the energy will be. The winding energy therefore has the form

$$E_2 = nR \qquad n = 1, 2, \ldots \tag{6.3}$$

where n is the so-called *winding number*. Combining equations (6.2) and (6.3), we can conclude that a wound up string has a total energy that can be expressed as

$$E = \frac{k}{R} + nR \tag{6.4}$$

What are the implications of this result? To see that, let us temporarily forget that energy is quantized, and assume that it has the form of a continuous function of R

$$E(R) = \frac{a}{R} + bR \tag{6.5}$$

In that case there is a minimal permissible energy, which corresponds to

$$R_{\min} = \sqrt{a/b} \tag{6.6}$$

Since energy is proportional to mass by virtue of $E = mc^2$, this also means that the mass of a wound up string cannot be smaller that some fixed $m_{\min}$.

Returning now to the quantized form (6.4), we should observe that the total energy exhibits a special kind of symmetry. Indeed, given any state

$$E(k, n, R) = \frac{k}{R} + nR \tag{6.7}$$

there exists a corresponding state $E(\hat{k}, \hat{n}, \hat{R}) = E(k, n, R)$, where $\hat{R} = 1/R$. This is easily verified by setting $\hat{n} = k$ and $\hat{k} = n$. Among other things, this implies that for every energy state with a given R there is an identical state with $\hat{R} = 1/R$. Consequently, states arising in very *large* spaces have exact counterparts in very *small* spaces.

This kind of reasoning leads us to the conclusion that all physical processes in a very thin cylindrical universe in which $R < 1$ (measured in units of Planck length) are identical to physical processes with $\hat{R} > 1$. In such a model, there is no "Big Crunch" in which the universe shrinks to the size of a point. Instead, when the size of the universe decreases to a certain nonzero radius R_0, any further reduction results in phenomena that are *exactly* the same as if the space were expanding. This effect is sometimes referred to as the *cosmic bounce*.

Black Hole Entropy

Entropy can be defined in terms of the number of microscopic states that produce the same macro-state. In a state of high entropy, almost all microstates correspond to the same macrostate. This implies a high degree of disorder, in which individual microstates are virtually indistinguishable. In the early 1970s, it was suggested that there is a connection between the surface area of a black hole's event horizon and entropy. The reasoning behind this conjecture was based on the recognition that entropy increases in every interaction (by the second law of thermodynamics), just like the area of the event horizon increases whenever the black hole interacts with an object.

There were a number of serious objections to this hypothesis (at the time, many physicists actually viewed it as no more than an analogy). The main difficulty was that wherever there is entropy, there must also be a non-zero temperature. This would imply a certain level of radiation from the black hole, which was in direct opposition to the claims of general relativity (according to which *nothing* can leave a black hole). In 1974, however, Stephen Hawking[5] showed that black holes are *not* completely black. He argued that the "quantum frenzy" which momentarily produces particle - antiparticle pairs exists everywhere, including the vicinity of the event horizon. The gravitational pull of the black hole could potentially absorb one of the particles into its interior, thus leaving the other one uncompensated. To an external observer, these uncompensated particles would then appear as radiation emanating from the black hole.

What is particularly interesting about Hawking's result (which involves some very complicated mathematics) is the fact that he managed to successfully combine quantum mechanics and general relativity. We should note, however, that although this theory did shed some light on the origin of black hole radiation, it provided no explanation whatsoever for the connection between the surface area of the event horizon and entropy. Some recent results of string theory propose a solution to this puzzle.

The conventional mechanism for the emergence of a black hole is the collapse of a massive star. When its nuclear fuel runs out, the star's outward pressure cannot counter the gravitational pull, and its entire mass is crushed into an extremely small space. Strominger and Vafa[6] proposed, however, that there could be an altogether different way in which black holes are formed. Basically, their approach amounted to carefully combining various "branes" that arise from M-theory, in order to ultimately produce a black hole. It turned out that many different "brane" combinations correspond to the same macroscopic state of the black hole. These combinations could be counted, and directly related to the definition of entropy. It was found that the entropy calculated in this way actually matches perfectly with the area of the black hole horizon as calculated by Hawking.

Other Implications

Among the many other potential implications of string theory, we should note that it allows for "tears" in the fabric of spacetime. Intuitively, one would expect that an event of this type would have catastrophic consequences for the universe. It turns out, however, that a moving string could "encircle" the tear in a way that resembles a protective cylinder, which effectively isolates such a tear until it "repairs" itself.

An even more interesting speculation concerns the very nature of spacetime. Just like an electromagnetic field can be viewed as a "sea" of photons, a gravitational field can be represented as a vast collection of gravitons (the fact that they have not been discovered experimentally doesn't seem to bother too many physicists). According to the principles of string theory, gravitons are just strings vibrating with a particular resonant pattern. With that in mind, it is possible to view spacetime itself as a "sea" of strings vibrating in a specific coherent manner.

Such a description naturally leads us to inquire whether there existed a primeval scenario where the constituent strings did not yet vibrate in a coherent manner. If this were possible, we would be dealing with circumstances in which spacetime has not yet been formed. This would be some sort of spaceless and timeless environment in which strings exist, but have not begun to vibrate with the coherent pattern needed to produce time and space. In other words, the "raw materials" for the creation of spacetime would be in place, but spacetime itself would not exist. This, of course, is a highly speculative claim, but it is being considered as a rational possibility in the context of M-theory.

Weaknesses of String Theory

One of the most serious weakness of string theory is the lack of experimental confirmation. Physicist Roger Penrose provides the following assessment of the underlying difficulties:

> "Many of the string theorist's claims are strongly asserted with apparent confidence. Undoubtedly, these must be watered down and taken with a sizeable heap of salt before serious consumption is contemplated. I think that it is fair to say that some of the strongest claims can be discounted altogether (such as their having provided a complete and consistent theory of quantum gravity). But having said that, I have to admit to there being the appearance of something of genuine significance 'going on behind the scenes' in some aspects of string/M-theory." [7]

Can string theory hope to produce any evidence to counter such criticism? Prominent string theorist Brian Greene [8] outlines a number of possibilities, three of which we note in the following:

1) Since string theory incorporates supersymmetry, it necessarily requires the existence of superpartners for each known particle. It is conceivable that one of these may be discovered some day, perhaps even in the near future. We should note, however, that this would not be a validation of string theory per se, but rather of the notion of supersymmetry. This concept is used in standard particle physics as well, and is not exclusive to string theory.

2) String theory allows for vibrational patterns that correspond to *fractionally charged* particles. If such charges were to be experimentally detected, this would be the validation of a prediction that is unique to string theory (nothing in standard physics suggests the existence of such particles).

3) The disintegration of protons is prohibited by the standard model of particle physics, but is permitted in the context of string theory. If such a process were to be discovered, it would constitute circumstantial evidence in favor of string theory.

Despite Greene's moderate optimism, string theory still faces stiff opposition in the scientific community. Nobel Prize winning physicist Sheldon Glashow and his colleague Paul Ginsparg have argued, for example, that abstract mathematical considerations cannot replace experimental verification as a method for establishing truth.

> "In lieu of the traditional confrontation between theory and experiment, superstring theorists pursue an inner harmony, where elegance, uniqueness and beauty define truth. The theory depends for its existence upon magical coincidences, miraculous cancellations and relations among seemingly unrelated (and possibly undiscovered) fields of mathematics. Are these properties reasons to accept the reality of superstrings? Do mathematics and aesthetics supplant and transcend mere experiment? ... For the first time since the Dark Ages, we

can see how our noble search may end, with faith replacing science once again."[9]

To secular scientists like Glashow and Ginsparg, the prospect that science may return (at least in some measure) to its speculative origins is clearly a negative development. From a theological perspective, however, this appears to be a very intriguing possibility. Among other things, it indicates that observation and repeated experiments may not be an adequate framework for investigating certain aspects of physical reality. Indeed, if string theory is correct in assuming that there are six additional spatial dimensions, there is no chance that we will ever be able to detect them (given their extremely small size). The same holds true for a number of other claims, whose validation would require unimaginably high levels of energy. The fact that science takes such propositions seriously (although they are essentially unprovable) suggests that we might consider allowing the same "freedom of interpretation" to theology.

6.3   Notes

1. Quoted in: Brian Greene, *The Elegant Universe*.

2. For further details, see *e.g.*: John Wheeler and Kenneth Ford, *Geons, Black Holes and Quantum Foam: A Life in Physics*, W. W. Norton, 1998.

3. It should be noted that each family contains an electron-like particle, two quarks and a neutrino-like particle. At the present time, the neutrino masses are not known precisely, and we only have upper bounds for them.

4. In 1957, mathematician Eugenio Calabi conjectured that a certain class of manifolds admits a Ricci-flat metric (a proposition that was subsequently proved by Shing-Tung Yau in 1977). In honor of this discovery, the family of such geometric objects bears the name Calabi-Yau spaces.

5. Stephen Hawking, *A Brief History of Time*, Bantam Books, 1996.

6. A. Strominger and C. Vafa, "Microscopic origin of the Bekenstein-Hawking entropy," *Physics Letters* B, **379**, pp. 99-104, 1996.

7. Roger Penrose, *The Road to Reality*, Knopf, 2004.

8. Brian Greene, *The Elegant Universe*.

9. P. Ginsparg and S. Glashow, "Desperately Seeking Superstrings?", *Physics Today*, **39**, pp. 7-9, May 1986.

Part II

The True, the Good and the Beautiful

Chapter 7

Aesthetics, Science and Theology

"The most beautiful thing we can experience is the mysterious. It is the source of all true art and science." *Albert Einstein* [1]

"Poets say science takes away from the beauty of the stars - mere globs of gas atoms. I too can see the stars on a desert night, and feel them. But do I see less or more? ... It does no harm to the mystery to know a little about it. For far more marvelous is the truth than any artist of the past imagined it." *Richard Feynman* [2]

The various theories and phenomena described in the preceding chapters provide us with a powerful conceptual framework for examining religious teachings from a scientific perspective. Before taking on this task, however, it might be helpful to identify certain fundamental questions that are of interest to both science and theology, and compare how these two disciplines approach them. Aesthetic theory appears to be a natural candidate for this type of analysis, since beauty plays a role in virtually every domain of human activity. With that in mind, in the following sections we will consider three important questions that relate to the nature and origins of beauty: the source of the aesthetic drive in humans, the relationship between truth and beauty, and the role that aesthetic criteria play in mathematics and physics. A closer examination of these topics will expose certain profound differences between the scientific and theological approaches. We will attempt to show, however, that these approaches needn't be mutually exclusive, and that the subtle connection between truth and beauty represents a common (and possibly unifying) theme for the two disciplines.

7.1   The Origins of the Aesthetic Drive

The notion that our sense of beauty has its roots in biology finds considerable support in the scientific community. Those who argue in favor of this view often point to the fact that our standards for judging human beauty have a strong utilitarian component, since they are conditioned by the sexual instinct. It has also been suggested that our preference for symmetry may have grown out of the need to distinguish between living creatures and inanimate objects, which was of considerable value to our ancestors. Such arguments indicate that our aesthetic drive had very practical origins, which have more to do with the survival of the species than with the creation of beautiful forms. Certain biologists, however, have taken a far more radical view, and have claimed that beauty has *no value at all* outside the context of evolution.

> "Beauty is not a feature of the world that has its own effect on us; instead, it is a projection of the emotions that we feel for merely Darwinian reasons. Even our tendency to believe that beauty is a causally important power, an objectively real datum, is engineered by genes that get an 'advantage' from our fond delusion." *E. O. Wilson* [3]

Wilson's statement may sound harsh, but it is by no means an isolated opinion. There is, in fact, an entire subdiscipline of psychology whose primary focus is the study of the aesthetic drive in humans. Some theorists in this field have associated aesthetic appeal with the mating ritual, and consider it to be a measure of biological fitness. Others have viewed art as a tool for solving cognitive problems and abstracting essential features of objects, or as a way to promote social cohesion. Yet another school of thought suggests that our ancestors developed aesthetic sensibilities as a way of coping with fear and uncertainty. Proponents of this last approach point to anxiety-reducing collective rituals such as chanting or dancing, which served a specific social purpose before they evolved into genuine aesthetic forms. [4]

It goes without saying, of course, that such theories have little in common with the theological view that beauty is first and foremost a reflection of its divine source. This interpretation, which has its roots in the earliest Christian texts, has always played a central role in the religious understanding of reality. [5] It is fair to say, however, that its influence has gradually diminished over the years, under pressure from an increasingly secular society. In describing the challenges that theological aesthetics faces today, Alejandro Garcia-Rivera singles out the fact that the unity of Beauty and the beautiful, which was once almost implicitly understood, now rarely arises as a topic in philosophical debates.

> "We have lost confidence, perhaps belief, in the human capacity to know and love God as Beauty. Thus, while some may still believe that God is the source of Beauty, and many that the beautiful can be experienced, few would be willing to say that these two are connected in a profound and organic way." [6]

For those who subscribe to this point of view, the obvious dilemma is whether one can reasonably believe that our sense of beauty is *both* a product of evolution *and* a reflection of God's essence. Although these two propositions appear to be at odds with each other, it is not particularly difficult to see that they are not mutually exclusive. Indeed, it is quite possible that music, dance and painting initially appeared as behavioral adaptations, whose primary purpose was the well-being and survival of primitive societies. Over the past few millennia, however, the various forms of art have become increasingly more sophisticated, with little or no practical value associated with them. This development has by no means diminished the status of art in society, or the universal appreciation that it commands – quite the contrary, in fact. Humanity has actually come to a point where the admiration of beauty can sometimes approach the kind of 'disinterested love' that is often described by religious mystics. Such a sentiment appears to have nothing to do with ownership or utility, and even less with the survival of the species. From that perspective, one could legitimately argue that our sense of beauty has transcended its original biological function, and has developed into something qualitatively different.

> "The reason dinosaurs once developed feathers (we may guess) was that they gave an advantage for temperature control. Having feathers, their descendants flew. ... Biologically, we have, as it were, our 'feathers' because – in devious and unexpected ways – they made our ancestors 'fitter'. Because we have them, we can – on occasion 'fly'." *Stephen Clark* [7]

It is, of course, difficult to say whether this kind of development ultimately serves a "higher" spiritual purpose (as some theologians have suggested). This is clearly not a question that science is equipped to answer. What we can do, however, is evaluate whether such an assertion is consistent with the human experience. In order to do that, it might be helpful to take a closer look at how theologians explain our sense of beauty and its origins. One of the most interesting interpretations of this kind relates to the claim that God made us in His likeness, and that the purpose of our existence is to realize this potential. If we were to accept such a premise, it would be perfectly reasonable to expect that man should be a creator by nature, and that the aesthetic drive provides a way to satisfy such a need. There would be nothing inherently unscientific or inconsistent in such an outlook, since it allows for the possibility that our creative tendencies could have developed as a natural outgrowth of evolutionary processes (see Section 10.3 for a more thorough discussion on this subject).

From a theological standpoint, this is clearly a promising approach, since it provides a natural connection between the religious notion of beauty and human creativity. We should keep in mind, however, that such a theory has its share of difficulties. A number of thinkers have pointed out, for example, that very few among us are as talented as Mozart, Einstein or Michelangelo. If we agree that this is the true standard of creativity, then theologians must explain why a loving God would bestow this gift so selectively, and deprive so many people of the possibility to realize the true purpose of their existence.

One could respond to such criticism in two different ways. The first would be to point out that human creativity has multiple levels, and that it is inappropriate to recognize only the highest one. On this view, each one of us has a genuine opportunity to create, and beauty can be found (at least, in some measure) in every human activity. What ultimately matters is the *creative process*, and not the quality of the final product.

A somewhat different approach would be to argue that the *appreciation* of beauty contains a genuine element of creativity. Indeed, when we comprehend a subtle point of Einstein's theory, or perhaps grasp the complex structure of Bach's fugues, the predominant sensation is undoubtedly one of profound satisfaction. Our "eureka" may not be completely original, but the enlightenment that we experience in the process may well be comparable to the one experienced by the author. Nobel Prize winning physicist S. Chandrasekhar openly endorsed such a view:

> "It does not follow that beauty is experienced only in the context of great ideas and by great minds. This is no more true than that the joys of creativity are restricted to a fortunate few. They are, indeed, accessible to each one of us provided we are attuned to the perception of ... the conformity of the parts to one another and to the whole." *S. Chandrasekhar* [8]

If we accept this interpretation, it is reasonable to believe that a focused *perception* of beauty can help elevate one's spirituality in much the same way as the creative process itself. Such a state of deep aesthetic appreciation can be a profound and almost mystical experience, which suggests (perhaps even to skeptics) that there is more to reality than meets the eye. Contemporary theologian Richard Viladesau speculates that this is precisely how art "opens us up" to the possibility of self-transcendence, which is a necessary element in any religious experience.

> "Clearly, it is not necessary to have Christian faith to appreciate the skill and beauty of Bach's music, apart from its message, and it would be possible to enjoy such a cantata on one level by consciously abstracting from the text. But to do so would deprive the experience of a major part of its aesthetic power. If one is really to feel the beauty of the emotion Bach evokes, one must attempt to feel, at least for a moment, what he felt, and one must therefore in some sense believe with him in the ultimate beauty he is presenting. ... For the act of aesthetic appreciation to take place, there must be a willing, if only momentary, suspension of disbelief: a willingness to see life, at least for this moment *as if* the hearer shared Bach's faith. This aesthetic act is of course far from real assent; yet it is significant, for it involves a certain openness to faith as a genuine human possibility." *Richard Viladesau* [9]

Philosopher Roger Scruton holds a similar opinion:

> "The experience of beauty points us beyond this world, to a 'kingdom of ends' in which our immortal longings and our desire for perfection are finally answered. As Plato and Kant both saw, the feeling for beauty is proximate to the religious frame of mind, arising from a

humble sense of living with imperfections, while aspiring towards the highest unity with the transcendental." *Roger Scruton* [10]

It is interesting to note in this context that the views expressed by Viladesau and Scruton have a long history, and can be traced back to the writings of medieval scholars. A typical representative of this school of thought was the 12th century statesman and patron of the arts, Abbot Suger, who claimed that the beauty of artistic objects has a transcendent quality, which can potentially elevate the human spirit into a "higher" mode of being. He summarized his own experience in the following way:

> "When - out of my delight in the beauty of the house of God - the loveliness of the many-colored gems has called me away from external cares ... it seems to me that I see myself dwelling, as it were, in some strange region of the universe which neither exists entirely in the slime of the earth nor entirely in the purity of Heaven; and that, by the grace of God, I can be transported from this inferior to that higher world." [11]

The idea that human creativity and aesthetic appreciation may have a spiritual dimension (despite their apparent biological origins) is not the only way to reconcile the scientific and religious interpretations of beauty. An alternative approach might be based on the theories of Carl Jung [12], who distinguished between four states of consciousness:

The surface mind, where our conscious mental activity takes place.

The preconscious, where our recent memories are stored, and from which they can be recalled at will.

The personal unconscious, where a wide array of forgotten memories is deposited.

The collective unconscious, which is the source of much of our instinctive behavior.

According to Jung, the collective unconscious is the deepest level of consciousness, shaped by the common experience of the entire human race. It is characterized by primordial images or *archetypes*, which are often encountered in myths and religious symbols. In his later writings, Jung likened the collective unconscious to an elaborate system of roots, which sustains and ultimately transcends individual life:

> "Life has always seemed to me like a plant that lives on its rhizome. Its true life is invisible, hidden in the rhizome. The part that appears above ground lasts only a single summer. Then it withers – an ephemeral apparition. When we think of the unending growth and decay of life and civilization, we cannot escape the impression of absolute nullity. Yet I have never lost a sense of something that lives and endures underneath the eternal flux. What we see is the blossom, which passes. The rhizome remains." *Carl Jung* [13]

Jung himself was a strict empiricist, and was reluctant to give such statements an explicit religious connotation. It is not difficult to recognize, however, that his theories naturally lend themselves to theological interpretations.

> "The conscious mind is like the one-eighth of an iceberg which appears above the surface. The seven-eighths not apparent to the naked eye is the subconscious. Beyond the subconscious is the collective unconscious, the noosphere, universal consciousness, or the total Godhead. This collective unconscious is like the ocean in which all the icebergs float. This ocean is not only the means of communication between all the icebergs, but is the collection of total truth to which we, as individual icebergs, have access." *Diarmuid O'Murchu* [14]

Following Jung's model, mathematician H. E. Huntley [15] proposed that the aesthetic experience is most powerful when it appeals to the collective unconscious. In its theological version, this argument maintains that the deepest level of our consciousness is rooted in the divine mystery that lies at the heart of all reality. From that perspective, our appreciation of beauty can be viewed as a way of learning something about this mystery, despite the fact that its essence remains beyond human comprehension.

> "Mystery, as theologians well know, both reveals and conceals. As such, mystery is not ignorance. Neither would mystery lead us to silence as Wittgenstein's maxim in Tractatus no. 7 suggests: "what we cannot speak about, we must be silent about." Something, or rather Someone becomes known in the mystery even as mystery makes us realize how little we know. Similarly, the beautiful allows Beauty to be felt even as Beauty itself slips, in the end, past the grasp of our affection." *Alejandro Garcia-Rivera* [16]

What conclusions can we draw from the preceding discussion? Both of the theological positions that we have outlined in this section allow for the possibility that our aesthetic preferences originally developed as adaptations to collective living. In that respect, they are certainly consistent with the theory of evolution. The only real source of controversy in this debate is the claim that our sense of beauty is an acquired trait whose *exclusive* purpose is the survival of the species. That is clearly not something that theologians can accept. It seems to me, however, that this particular discrepancy doesn't warrant an extended discussion, since purely biological explanations of our aesthetic preferences tend to be rather one-sided and simplistic. Among other things, they fail to account for the fact that complex systems often give rise to emergent phenomena, which *include* and *transcend* the simpler forms that they grew out of. Given that the human mind is arguably the most complex system that we know of, it is perfectly reasonable to assume that emergence may have played a key role in the development of our aesthetic drive. If we allow for such a possibility, we can plausibly argue that our sense of beauty has outgrown its original biological function, and that this faculty can perhaps guide us to a kind of truth that cannot be grasped by analytic thinking alone.

7.2 Beauty and Truth

The idea that beauty and truth are intimately related is not exclusive to theologians, and has been embraced by many scientists as well. One of the most notable proponents of this view was the famous physicist Paul Dirac, who once remarked that: "It is more important to have beauty in one's equations than to have them fit experiment" [17]. This strange and counterintuitive statement raises a number of intriguing questions about the role that aesthetic criteria play in the domain of science. Our objective in the following will be to examine some of these questions in greater detail, and provide several illustrative examples.

We begin with two preliminary observations, the first of which concerns the difference between theory and experiment. Although scientists generally place a high value on data gathering and practical applications, many of them have passionately argued that theory is worth pursuing for its own sake. Such a view is often justified by the fact that theories are designed to give form to formless things, which has a clear aesthetic appeal. In that respect, artists, physicists and mathematicians seem to have a lot in common.

"The work is not done for the sake of an application. It is done for the excitement of what is found. ... Without understanding this, you miss the whole point." *Richard Feynman*[18]

"In theory, there is no difference between theory and practice. But, in practice, there is." *Yogi Berra*[19]

The second observation is that there are areas of modern science and mathematics which are very abstract, and have little or no connection with human experience. In quantum mechanics and string theory, for example, there are no clear empirical guidelines for identifying which mathematical models are appropriate and which are not. Under such circumstances, scientists have often resorted to purely aesthetic criteria, and Dirac's statement is a reflection of that fact. The following example shows that this line of reasoning is certainly not divorced from reality.

Example 7.1. The second half of the 20th century witnessed the discovery of a wide array of subatomic particles. While many of them were detected experimentally, some were actually identified by purely aesthetic considerations. A classic example is the prediction of the so-called Ω^- in 1962. In analyzing a class of heavy particles, Murray Gell-Mann and his associates plotted two quantities known as *hypercharge* and *isospin*, obtaining the diagram shown in Fig. 7.1. It turned out that each vertex in this diagram corresponded to a known particle *except* for the bottom one. Inspired by the inherent symmetry of the graph, Gell-Mann anticipated that there ought to be a particle associated with the empty spot as well. Not surprisingly, several years later the missing Ω^- particle was detected experimentally.

It is important to keep in mind that Gell-Mann's result is just one of many scientific discoveries that were based on aesthetic considerations. The idea that beauty and scientific truth are intimately related is actually very old, and can

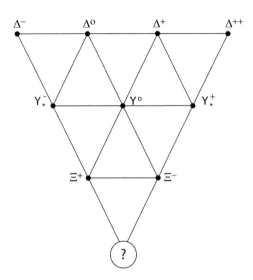

Fig. 7.1. Gell-Mann's diagram with the missing particle.

be traced back to ancient Greece. One of the most notable proponents of this approach was Pythagoras, whose view of nature was greatly influenced by the notions of beauty and harmony. He suggested (among other things) that each of the seven planets that were known in his day produced a particular note, whose pitch depended on its distance from earth. He believed that the collective motion of the planets gave rise to a "music of the spheres," which was inaudible to humans. Throughout the middle ages, this "music" was thought to be the most beautiful combination of sounds in the universe, and the sign of an ultimate cosmic order.

Needless to say, this type of reasoning has little to do with science as we know it today. And yet, it indirectly lead to one of the most important discoveries in the history of science. To see how that came about, we should recognize that astronomers of the early Renaissance were still heavily influenced by the theories of Pythagoras. They were, however, increasingly bothered by the fact that planets produced only seven different notes, which violated their sense of aesthetic perfection (since the complete scale consists of eight tones). The only way to resolve this problem was to assume that the earth moves as well, and produces a note of its own. This bold hypothesis undermined the Ptolemaic model of the heavens, and ultimately opened the door to the Copernican Revolution.

Although this particular example is mainly of anecdotal interest, it does highlight a principle that has always played an important role in the history of human thought. Indeed, even when science began to clearly differentiate itself from philosophy and theology, its practitioners continued to view aesthetic criteria as a guide to the truth. It is well known, for instance, that the notion of symmetry inspired Newton to formulate his law of action and reaction, and that similar considerations lead Mendeleev to anticipate unknown elements in the periodic system. The list goes on and on, with modern particle physics and unification theories as the most recent examples.

It is important to recognize in this context that symmetry and order are not the only important aesthetic concepts in nature - *symmetry breaking* is probably just as relevant. This notion plays a key role in particle physics and cosmology, and suggests that genuine novelty in nature requires a certain amount of irregularity in its basic structures. Indeed, had the amounts of matter and antimatter been identical in the aftermath of the Big Bang, the two would have cancelled each other out and the universe as we know it would never have emerged. Symmetry-breaking is also responsible for the existence of four fundamental forces in nature (instead of a single unified "superforce"). The fact that these forces have very different characteristics is an essential prerequisite for the formation of atoms and molecules.

The delicate balance between symmetry and asymmetry in nature has an aesthetic appeal of its own. In that respect, science is not unlike the game of chess – the rules appear to be simple and symmetric, but the game itself offers an infinite variety of forms. In both cases, the rules provide a way for establishing unity and order in the midst of diversity. What has fascinated researchers most in this context has been the fact that aesthetic principles often lead to completely unanticipated scientific discoveries. We can legitimately wonder why mathematical beauty (which is essentially a subjective category of the human mind) turns out to be an effective tool for describing and predicting the objective reality of nature. There is an element of profound mystery in this question.

"The eternal mystery of the world is its comprehensibility." *Albert Einstein* [20]

"There is no a priori reason why beautiful equations should prove to be a clue to understanding nature. ... I believe that the rational beauty of the cosmos indeed reflects the Mind that holds it in being." *John Polkinghorne* [21]

Perhaps part of the answer lies in the implicit connection between *knowledge, beauty* and *simplicity*. Most scientists will agree, for example, that very complicated theories are seldom viewed as "beautiful," and are often an indication of randomness rather than lawful behavior. Simple theories, on the other hand, attract us both aesthetically and practically, since physical observations that are reducible to a rudimentary pattern can usually be explained in an elegant manner. [22] This link between simplicity, beauty and scientific explanations gives rise to some interesting theological speculations. One such interpretation suggests that simplicity is aesthetically appealing to us *precisely* because it is a fundamental characteristic of God's creation, and a key to understanding it. Those who subscribe to this point of view hold that our ability to comprehend reality is a part of the Divine purpose. This purpose obviously couldn't be fulfilled unless a significant proportion of complex phenomena were reducible to elementary laws.

"God has chosen that which is the most perfect, that is to say, in which at the same time the hypotheses are as simple as possible, and the phenomena are as rich as possible." *Gottfried Wilhelm von Leibniz* [23]

Leibniz's claim that we live in the "best of all possible worlds" is based on two fundamental theological assumptions about nature and our place in it. The first one is that the universe is *intelligible*, and that many of its laws are simple enough to be grasped by the human mind. Such a belief is central to the scientific enterprise, since it motivates us to search for regularities in the data that we observe. On this point, Einstein and Leibniz seem to be in perfect agreement:

> "A conviction, akin to religious feeling, of the rationality or intelligibility of the world lies behind all scientific work of high order."
> *Albert Einstein* [24]

The second assumption that is implicit in Leibniz's thought is that the universe is *contingent*. He believed (as do many theologians) that the laws of nature are a result of God's *choice*, and that this particular form of organization was not logically necessary. The possibility that things could have turned out differently suggests that creation may have a higher purpose, which is gradually realized through the emergence of a wide variety of natural forms. Such a view finds support in the fact that this process ultimately gave rise to complex forms of organization such as the human mind, which allow the universe to "contemplate itself" (and perhaps its creator as well).

Regardless of what we may think of such arguments, I believe we can agree that there is something truly remarkable in the way nature appears to be structured. The fact that such a broad range of phenomena can result from a small number of relatively simple laws represents a kind of "unity in diversity" that is aesthetically appealing to scientists and theologians alike. Our predisposition to look for such unifying schemes has lead to many unexpected discoveries about the physical world, particularly in areas where the possibilities for experimental testing are limited.

The connection between truth and beauty can be made along somewhat different lines, by observing that art, physics and theology deal with aspects of reality for which no appropriate words can be found. In his book *Art and Physics*, Leonard Shlain [25] argues that artists often intuitively sense new concepts, and express them symbolically *before* scientists can articulate them in terms of mathematical models. According to this view, revolutionary artistic ideas are often the first stage of a process in which mankind begins to alter its perception of the physical world, and uncovers previously hidden relationships. This is an intriguing proposition, which finds support among scientists, theologians and poets alike.

> "Both science and art form in the course of the centuries a human language by which we can speak of the more remote parts of reality."
> *Werner Heisenberg* [26]

"Great art can communicate before it is understood. *T. S. Eliot* [27]

Shlain's book cites numerous examples of artistic work that anticipated ground breaking scientific discoveries about space, time and light. He claims, for instance, that the concept of "curved" space made its first appearance in the paintings of Edouard Manet, some fifty years before Einstein formulated the theory of general relativity. Shlain points out that Manet routinely distorted "linear" objects such as trees and hats, and even curved the horizon in some cases. Such paintings clearly challenged conventional perceptions of physical reality, and met with harsh criticism at the time. It turned out, however, that Manet's intuition was correct, and by the 1920s it was experimentally verified that physical space is indeed "curved" (see Chapter 5 for a detailed discussion). According to Shlain, such examples underscore the connection between what we perceive as beautiful, and what turns out to be true. From that perspective, it is perfectly reasonable to think of science and art as different projections of the same reality, with beauty as a prominent common element.

> "What is beautiful in science is the same thing that is beautiful in Beethoven. There is a fog of events and suddenly you see a connection. It expresses a complex of human concerns, that goes deeply to you, that connects things that were always in you that were never put together before." *Victor Weisskopf* [28]

Weisskopf's statement is clearly consistent with the theological view that beauty and creativity are intimately related to spiritual growth, and can lead to a "higher" form of truth. In this process, the artist often plays the role of a "catalyst." A character in one of Aldous Huxley's stories eloquently describes what such a role entails:

> "The main function of art is to impart knowledge. The artist knows more than the rest of us. He is born knowing more about his soul than we know of ours, and more about the relations existing between his soul and the cosmos. He anticipates what will be common knowledge in a higher state of development. ... Moreover, ... he can say what he knows, and say it in such a way that our own rudimentary, incoherent, unrealized knowledge of what he talks about falls into a kind of pattern – like iron fillings under the influence of the magnet." [29]

Although aesthetic experiences of this kind to tend to be deeply subjective, that doesn't automatically make them any less "real" or relevant. Indeed, even within physics itself, the distinction between subjectivity and objectivity is far from clear:

> "The common division of the world into subject and object, inner and outer world, body and soul, is no longer adequate." *Werner Heisenberg* [30]

> "Even the laws of physics themselves may be somewhat observer dependent." *Stephen Hawking* [31]

The views expressed by Heisenberg and Hawking are of interest to us for a number of reasons. To begin with, they are surprisingly similar to the theological claim that reality has both an objective and a subjective dimension. They also suggest that these two domains overlap in subtle ways, and cannot be completely separated. One of the things that binds them together is the idea that purely subjective aesthetic considerations can serve as a guide to objective truth. This possibility is equally appealing to theologians and scientists, and indicates that truth and beauty may just be different ways of characterizing the same overarching reality. Perhaps the great poet John Keats sensed this connection when he wrote the famous lines:

"Beauty is truth, truth beauty, that is all

Ye know on earth, and all ye need to know." [32]

We must take care, however, not to take this line of reasoning too far. Beauty may indeed be a guide to the truth, but it would be a mistake to use it as the *only* relevant criterion. As Nobel Prize winning economist Paul Krugman points out, such an outlook can sometimes be thoroughly misleading:

"As I see it, the economics profession went astray because economists, as a group, mistook beauty, clad in impressive-looking mathematics, for truth." *Paul Krugman* [33]

7.3   Beauty in Mathematics and Science

If theologians are correct in assuming that beauty is a guide to the truth, then one would expect to find evidence of this in all areas of human activity, including math and science. We already touched on this issue in Section 7.2, and considered several examples that support such a claim. We now need to expand this analysis and examine a number of other aesthetic criteria that are relevant in this context.

The Creation of Form

There is clearly a kind of illumination and relief that accompanies the solution of a scientific problem that has puzzled us for a long time. This feeling is more than just a sense of skill and accomplishment – it is also associated with the creation of order out of disorder.

"The glory of science is not in a truth more abstract than the truth of Bach and Tolstoy, but in the act of creation itself. The scientist's discoveries impose his own order on chaos, as the composer or painter imposes his; an order that always refers to limited aspects of reality, and is based on the observer's frame of reference, which differs from period to period as a Rembrandt nude differs from a nude by Manet." *Arthur Koestler* [34]

The following two examples illustrate how scientific order can be entirely abstract in some cases, while in others it has a more direct visual interpretation.

Example 7.2. (Power Systems) Electric power systems are usually described as a collection of generators and loads, which are connected by a network of transmission lines. A schematic representation of such a network has the form of a graph such as the one shown in Fig. 7.2.

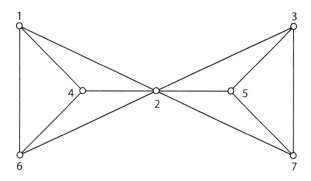

Fig. 7.2. A graph theoretic representation of a power system.

A graph like this can always be equivalently described in terms of an *incidence matrix* A, whose elements satisfy $a_{ij} \neq 0$ if and only if nodes i and j are directly connected. The incidence matrix for the graph in Fig. 7.2 is given in equation (7.1), where blank spaces represent zeros (the diagonal elements are assumed to be nonzero by convention).

$$
A = \quad
\begin{array}{c}
 \\ 1 \\ 2 \\ 3 \\ 4 \\ 5 \\ 6 \\ 7
\end{array}
\begin{array}{c}
\begin{array}{ccccccc}
1 & 2 & 3 & 4 & 5 & 6 & 7
\end{array} \\
\left[
\begin{array}{ccccccc}
* & * &   & * &   & * &   \\
* & * & * & * & * & * & * \\
  & * & * &   & * &   & * \\
* & * &   & * &   & * &   \\
  & * & * &   & * &   & * \\
* & * &   & * &   & * &   \\
  & * & * &   & * &   & *
\end{array}
\right]
\end{array}
\qquad (7.1)
$$

The nonzero pattern of an incidence matrix is very important in circuit analysis and simulation. Since this pattern depends heavily on the way in which the nodes are numbered, it is of interest to examine whether there is an optimal way to do this. To get a sense for why this is important, suppose that we were to renumber the nodes in the manner shown in Fig. 7.3.

The resulting incidence matrix would then have the structure shown in (7.2), which is characterized by two 3×3 diagonal blocks and a "border" that corresponds to row and column 7. This type of structure is known as the Bordered Block Diagonal (BBD) form, which is well suited for parallel computation.

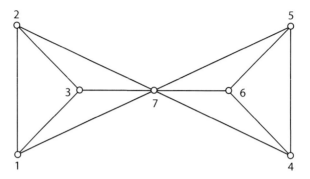

Fig. 7.3. The graph with renumbered nodes.

$$A = \begin{array}{c} \\ 1 \\ 2 \\ 3 \\ 4 \\ 5 \\ 6 \\ 7 \end{array} \begin{array}{ccccccc} 1 & 2 & 3 & 4 & 5 & 6 & 7 \\ \left[\begin{array}{ccccccc} * & * & * & & & & * \\ * & * & * & & & & * \\ * & * & * & & & & * \\ & & & * & * & * & * \\ & & & * & * & * & * \\ & & & * & * & * & * \\ * & * & * & * & * & * & * \end{array}\right] \end{array} \tag{7.2}$$

In realistic power systems, the number of nodes tends to be large, and the identification of an appropriate ordering becomes a very difficult problem. A typical example of such a situation is provided in Fig. 7.4, which represents a matrix that arises in the analysis of the Western U. S. power grid (the model consists of 5,300 nodes). This matrix exhibits no discernible pattern, since the nodes in the original graph were numbered arbitrarily. It turns out, however, that through a clever renumbering we can obtain the BBD structure shown in Fig. 7.5. The fact that we can do this is a source of aesthetic pleasure which is not directly related to the resulting computational benefits (although the synergy of the beautiful and useful clearly adds to the appeal).[35] Those who find any beauty in Fig. 7.5 will most likely derive this feeling from the recognition of a definite form, which emerges out of something previously formless.

Example 7.3. (The Prime Number Theorem) A considerably more abstract example of order arises in the context of prime numbers. A careful examination of these numbers shows no pattern whatsoever – they seem to be randomly scattered throughout the set of positive integers. And yet, there is a subtle regularity that was discovered by the great French mathematician Jacques Hadamard toward the end of the 19th century. To comprehend the nature of this regularity, let us denote the number of primes that are less than or equal to some number

n as $\pi(n)$. In that case, the following simple relationship holds

$$\lim_{n \to \infty} \frac{\pi(n)}{(n/\log n)} = 1 \qquad (7.3)$$

Equation (7.3) represents one of the most important results of modern mathematics, and is commonly referred to as the Prime Number Theorem. This theorem tells us that $\pi(n)$ actually behaves like $n/\log n$ when n is sufficiently large. A closer analysis of this result shows that this is a good approximation of $\pi(n)$ only in the limit (that is, when $n \to \infty$). Accuracy, however, is not the main point here. What is important for our purposes is that equation (7.3) is aesthetically pleasing to a mathematician, since it uncovers a higher level of order and form amidst apparent randomness.

Since we are discussing the aesthetic appeal of order, it might be interesting to make a small digression, and examine how art and mathematics compare when it comes to creating new forms. There is a widespread belief that art

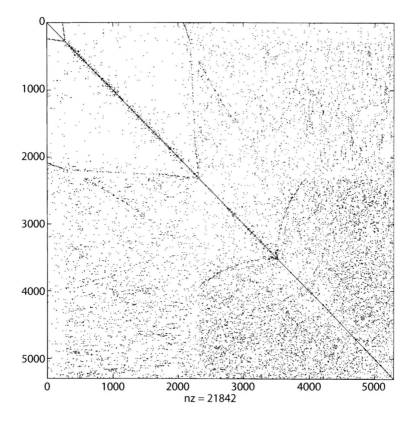

Fig. 7.4. The original matrix.

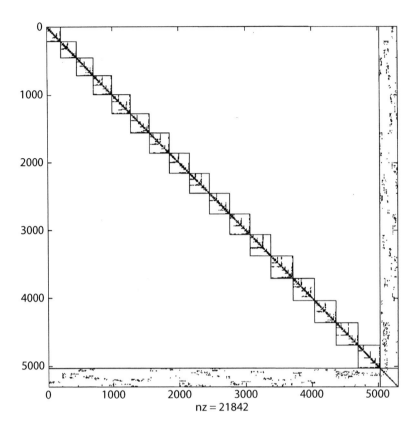

Fig. 7.5. The reordered matrix.

arises essentially out of 'nothing', through an act of individual inspiration. The 19th century Russian thinker Alexander Herzen clearly articulated this sentiment with the question: "Where was the painting before the artist painted it?" In contrast, many have held that mathematical forms are not created, but are rather *discovered*. This outlook (which can be traced back to Plato) is based on the belief that mathematical truths are absolute and eternal. Since the corresponding forms are in some sense 'waiting be found', mathematical innovation appears to be merely a matter of time and the right set of social conditions. On these grounds, one might legitimately wonder whether mathematical and artistic creativity can be placed on an equal footing.

    There are two ways to respond to this dilemma. One possible approach would be to directly challenge the neo-Platonic view of mathematics, as many philosophers of science have done in recent years. These thinkers maintain that we routinely impose our mathematical structures on the universe, in an attempt to describe its laws and regularities. The models that we obtain in the process may be more or less adequate reflections of reality, but they are never absolutely true. They are better described as projections of our thought patterns onto an objective external world. From this standpoint, mathematical creativity would appear to be very similar to its artistic counterpart – in both cases individual preferences and aesthetic criteria play a crucial role, and truth is only approximated.

An alternative approach would be to accept the belief that mathematics exists independently of the human mind, and that there are eternal truths to be found. In that case, the central question is whether the existence of such an 'objective' universe somehow diminishes the aesthetic value of scientific and mathematical discoveries. I would argue that it does not, despite the fact that logic and natural laws clearly place certain restrictions on what can be created. If we judge the scientific enterprise by the achievements of its greatest protagonists, we are bound to conclude that it is a highly creative activity, which leaves considerable room for personal interpretations.

> "Perhaps the best summary of the differences that are often perceived between science and arts is the following widely accepted statement: If you give ten scientists the same problem, they will reach exactly the same answer (assuming they each solve the same problem); but if you ask ten artists to paint the same scene, you will get ten different paintings. ... The notion that ten scientists will arrive at the same answer may describe what will happen when ten unimaginative scientists are asked to solve a problem for which the answer has already been published. It is not what one would expect if one gathered the Pasteurs, Darwins, Paulings, Einsteins, Bohrs, Woodwards and Feynmans of the world and asked them to address a problem."
> *Robert Root-Bernstein* [36]

The claim that mathematics and physics allow for individual expression finds considerable support in contemporary scientific practice. Indeed, in the natural sciences it is by no means unusual to encounter very different theoretical models for the same phenomenon, all of which are mathematically equivalent (in the sense that they yield identical predictions). This is true even of well established results such as the periodic table, which has over 400 equally "valid" versions (each of which represents a different way of organizing the element properties). [37] Such examples indicate that scientific data often allows for multiple interpretations, and suggest that creative researchers ought to embrace this ambiguity. In making this point, philosopher Robert Root-Bernstein cites the experience of two-time Nobel laureate Linus Pauling:

> "Pauling had written that what distinguished important from trivial science is the type of question one asks at the outset. Most scientists - average practitioners - ask, "What conclusions ... are we forced to accept by these results of experiment and observation?" Pauling says that he asks himself instead, "What ideas" - note the plural - "about this equation, as general and aesthetically satisfying as possible, can

we have that are not eliminated by these results of experiment and observation?" [38]

Pauling's words sound very much like something that a sculptor might say as he ponders the constraints imposed on him by the marble that he works with. In both cases there are certain restrictions regarding what is possible, but there is also a great deal of creative freedom. With that in mind, there is no good reason to assume that artists "create" while scientists merely "discover." It would be more appropriate to conclude (as Arthur Koestler did) that *both* sides are engaged in the process of discovery, each in its own way.

"Discovery often means simply the uncovering of something which has always been there but was hidden from the eye by the blinkers of habit. This equally applies to the discoveries of the [scientist and of the] artist who makes us see familiar objects and events in a strange, new, revealing light. ... Newton's apple and Cézanne's apple are discoveries more closely related than they seem." *Arthur Koestler* [39]

Minimalism

There is an obvious aesthetic quality in expressing the truth with a minimum of description, regardless of whether the medium is mathematics or poetry.

"You know you have achieved perfection in design not when there is nothing more to add, but when there is nothing left to take away." *Antoine de Saint-Exupery* [40]

Illustrations of such 'elegant' mathematical proofs are provided in Examples 7.4 and 7.5, both of which establish non-intuitive facts with extraordinary simplicity.

Example 7.4. The following result was obtained by the famous 19th century mathematician Georg Cantor. It states (and proves) a highly non-intuitive fact about infinite sets.

Theorem 7.1. The set of real numbers is *not* countable.

Proof. What follows is a *proof by contradiction*. As in all such cases, we will assume that the theorem is actually *incorrect*, and that we can form an infinite list in which each real number $r(n)$ is uniquely matched with a positive integer n (this is what the term 'countable' means). If we focus on the interval $[0, 1]$ for simplicity, our list could look something like the one shown in Table 7.1 (note that real numbers are represented by infinitely many decimals in general).

Our objective will now be to form a number that is *not* a member of the list. If we succeed, this will obviously contradict our initial assumption that the set of real numbers is countable. To construct such a number, let us subtract 1 from the diagonal (boldfaced) entry in each $r(i)$ ($i = 1, 2, \ldots$), with the understanding that subtracting 1 from 0 produces 9. The number obtained by assembling the modified diagonal terms would then be

$$x = .27509\ldots \tag{7.4}$$

$$r(1): \quad .\mathbf{3}1140\ldots$$

$$r(2): \quad .2\mathbf{8}433\ldots$$

$$r(3): \quad .52\mathbf{6}71\ldots$$

$$r(4): \quad .042\mathbf{1}0\ldots$$

$$r(5): \quad .0300\mathbf{0}\ldots$$

$$\vdots \qquad \vdots$$

Table 7.1. A hypothetical ordered list of real numbers.

By construction, x differs from $r(1)$ in the *first* digit, from $r(2)$ in the *second*, from $r(3)$ in the *third*, and so on. In other words, x is different from *every single number* in the list, and is therefore not a part of it. **Q.E.D.**

One of the major consequences of this theorem is the fact that it allows us to distinguish between different kinds of infinity. Essentially, it tells us that the set of real numbers is in some sense 'larger' than the set of integers, although both contain infinitely many elements. This is by no means obvious.

Example 7.5. Our second example of mathematical minimalism is a very old result, which was formulated by Euclid in the 4th century B.C. It is a remarkably simple proof of the fact that the set of prime numbers is infinitely large.

Theorem 7.2. There are infinitely many prime numbers.

Proof. Consider an arbitrary finite list of prime numbers, and let the largest number in the list be p. Our objective will be to show that there is always a prime number that is greater than p, regardless of how large p happens to be. In order to do that, let us form an integer

$$N = (2 \cdot 3 \cdot 5 \cdot \ldots \cdot p) + 1 \tag{7.5}$$

and consider the only two possible scenarios.

1. Suppose N is a prime number. In that case N is clearly larger than p construction.

2. Suppose N is *not* a prime number. If N is not prime, it can always be factored as a product of several prime numbers. However, from (7.5) we see that N cannot be divided by 2, 3, 5, $\ldots$, p without a remainder. As a result, if it really does have prime factors, then these numbers must be larger than p.

Both cases lead to the conclusion that there exists a prime number greater than p. Since p was chosen *arbitrarily*, it directly follows that the set of prime numbers is unlimited. **Q.E.D.**

In contrast to the previous two examples, let us now consider what a 'non-elegant' proof might look like.

Example 7.6. (The Coloring Problem) The Four Color Theorem is a graph-theoretic problem whose origins are quite practical. In 1852, mathematician Francis Guthrie was attempting to color a map of the counties of England, with the objective that no two neighboring counties should be assigned the same color. Such a map would obviously be easy to read, and the principle could be applied to a whole range of similar problems.[41] Guthrie's conjecture was that the minimal number of different colors needed for this task is four, but he was unable to prove the result rigorously.

The problem turned out to be extremely difficult, and the proof of this conjecture eluded mathematicians for over a century. In 1976, graph theorists Kenneth Appel and Wolfgang Haken finally managed to reduce the problem to a large (but finite) number of critical configurations that needed to be checked. It turned out that the only feasible way to do this was to use a computer, since the number of cases was far too large for manual verification.

Although the proof of this result was a major breakthrough, many mathematicians were thoroughly disappointed by the way in which it was obtained. The "brute force" method of counting and eliminating different possibilities lacked elegance, and offended the aesthetic sensibilities of those who find beauty in minimalist proofs.

> "When I heard that the Four Color Theorem had been proved, my first reaction was, "Wonderful! How did they do it?" I expected some brilliant insight, a proof which had in its kernel an idea whose beauty would transform my day. But when I received the answer, "They did it by breaking it down into thousands of cases, and then running them all on the computer, one after the other," I felt disheartened. My reaction was, "So it just goes to show it wasn't a good problem after all." *Philip Davis*[42]

Example 7.6 clearly suggests that simplicity and truth don't always go hand in hand, and that minimalism, while desirable, cannot become a universal requirement in science. In that respect, I think we can agree with Einstein that:

> "Things should be as simple as possible, but not simpler."[43]

Unity in Diversity

The desire to unify diverse phenomena has been a powerful motive throughout recorded history. Today there is virtually no domain of human knowledge without a theory that brings together seemingly unrelated ideas and observations. Scientific theories are especially interesting in this context, because their potential for unification is often supplemented with a capacity for prediction and validation. Such a harmonization of the true and the beautiful is undoubtedly one of the most profound sources of intellectual pleasure.

Unification plays a prominent role in modern physics, where numerous attempts have been made to identify an overarching symmetry that connects the basic forces of nature. This approach relies heavily on aesthetic criteria, which are often used to justify theories that cannot be verified experimentally. A typical example of this type is string theory, which is embraced by many scientists despite the fact that it has produced no empirical evidence to date. What it has produced, however, is an elegant theoretical framework that can potentially reconcile and unify quantum mechanics and general relativity (see Chapter 6 for more details).

It is of interest to note that unification is intimately related to the process of *abstraction*. This concept is as important in mathematics as it is in theology, since it provides a way of transcending the purely material aspects of the world.

> "Arithmetic has a very great and elevating effect, compelling the soul to reason about abstract numbers, and rebelling against the introduction of visible or tangible objects into the argument". *Plato* [44]

The unifying potential of mathematical abstraction is not difficult to see. When faced with a system of differential equations, for example, one cannot tell whether it originated from economics, biology or electric circuits. For mathematical purposes such distinctions are irrelevant, since the underlying models assume the same general form. A further illustration of this point is provided by the following example, which underscores the subtle connections that exist between trigonometric functions, exponentials and power series. [45]

Example 7.7. The notion of a power series makes its appearance very early in human history. Legend has it that an Indian king invited the inventor of the game of chess to his palace, in order to recognize the achievement. When asked to name his reward, the inventor made the following proposal: "One grain of wheat for the first square on the board, two for the second, four for the third, and so on until the sixty fourth square." The king was apparently pleased by the mathematician's modesty until he realized that

$$S = \sum_{n=0}^{63} 2^n \tag{7.6}$$

is actually a staggering amount (roughly 1.84×10^{19} grains). What happened to the inventor at this point remains unclear ...

Trigonometric ratios also appeared many centuries ago, but in a very different context. Their primary purpose was to formalize certain simple relationships between the sides and angles of a right triangle. Subsequent developments resulted in extensions of these concepts, and the introduction of more abstract trigonometric functions such as $\sin x$ and $\cos x$. For a very long time, these functions had absolutely nothing in common with power series or exponentials. All this changed in the 18th century, when an elegant unification was achieved in the form of Taylor series expansions. As is well known, these expansions allow us to express any *sufficiently smooth* function as a power series. Sines, cosines and exponentials satisfy this condition, and can therefore be expressed as

$$\sin x = x - \frac{x^3}{3!} + \frac{x^5}{5!} + \dots \tag{7.7}$$

$$\cos x = 1 - \frac{x^2}{2!} + \frac{x^4}{4!} - \dots \tag{7.8}$$

$$e^x = 1 + x + \frac{x^2}{2!} + \frac{x^3}{3!} + \dots \tag{7.9}$$

The similarity in form between the three expansions is evident at this level of abstraction, but the unification does not end there. If we replace x in (7.9) with ix (where $i = \sqrt{-1}$), it is easily verified that

$$e^{ix} = (1 - \frac{x^2}{2!} + \frac{x^4}{4!} - \dots) + i(x - \frac{x^3}{3!} + \frac{x^5}{5!} + \dots) \tag{7.10}$$

which leads to Euler's famous formula

$$e^{ix} = \cos x + i \sin x \tag{7.11}$$

Setting $x = \pi$, we further obtain

$$e^{i\pi} = -1 \tag{7.12}$$

Equation (7.12) is a fascinating example of the unifying power of mathematical abstraction. If we examine this expression closely, we recognize that it combines three of the most famous mathematical constants, π, e and i, into a *single* and extraordinarily simple relationship. Given that these constants had very different (and essentially unrelated) histories, (7.12) is truly a remarkable result whose aesthetic appeal is undeniable.

Unsuspected Relationships and Patterns

From an intellectual standpoint, the discovery of unanticipated relationships and patterns is one of the most rewarding experiences. It could be argued, for instance, that chaos theory is intriguing largely because it uncovers a surprising correlation between apparently random phenomena. This feeling is amplified when one encounters the fascinating geometric patterns that characterize strange attractors. In the following we will explore this aspect of aesthetic appeal more closely, in the context of the diverse and unexpected way in which the number $\phi = (1 + \sqrt{5})/2$ occurs in mathematics and nature.

Number ϕ makes its first historical appearance in antiquity, in relation to the so-called *Golden Section*. To see what this means, consider the right triangle shown in Fig. 7.6, where $AB = l$ and $BD = l/2$.

If we construct arcs BE and EC, it is not difficult to show that [46]

$$AC = \frac{l(\sqrt{5} - 1)}{2} \tag{7.13}$$

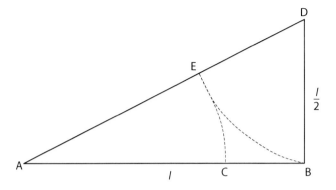

Fig. 7.6. An illustration of the Golden Section.

and

$$BC = AB - AC = \frac{l(3 - \sqrt{5})}{2} \tag{7.14}$$

As a result,

$$\frac{AB}{AC} = \frac{AC}{BC} = \phi. \tag{7.15}$$

This ratio fascinated the ancient Greeks to the point that they built it into the proportions of the Parthenon. The school of Pythagoras even attributed mystical properties to ϕ, since it appeared in many relationships that characterize the pentagon. A simple illustration is provided in Fig. 7.7, which represents a pentagon with sides of unit length. It can be shown that the diagonals intersect in such a way that

$$\frac{MP}{PW} = \frac{BP}{AP} = \phi. \tag{7.16}$$

Number ϕ shows up again in the 12th century, this time in the context of algebra. Leonardo of Pisa (also known as Fibonacci) studied sequences of numbers that were generated by the recursive formula

$$u_{n+1} = u_n + u_{n-1} \tag{7.17}$$

He found that no matter how the first two terms, u_0 and u_1, are chosen, the ratio u_{n+1}/u_n satisfies

$$\lim_{n \to \infty} \frac{u_{n+1}}{u_n} = \phi \tag{7.18}$$

The special case when $u_0 = 1$ and $u_1 = 1$ produces the sequence shown in Table 7.2, whose elements are referred to as *Fibonacci numbers*.

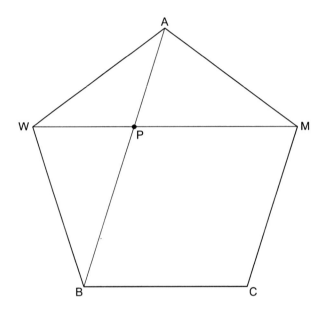

Fig. 7.7. Number ϕ and the pentagon.

$u_0 = 1$	$u_6 = 13$
$u_1 = 1$	$u_7 = 21$
$u_2 = 2$	$u_8 = 34$
$u_3 = 3$	$u_9 = 55$
$u_4 = 5$	$u_{10} = 89$
$u_5 = 8$	$u_{11} = 144$

Table 7.2. The first 12 Fibonacci numbers.

Of course, to those who are familiar with difference equations, property (7.18) comes as no great surprise. Indeed, it is not difficult to see that the solution of equation (7.18) must have the form

$$u_n = a_1\phi^n + a_2\left(\frac{1}{\phi}\right)^n \tag{7.19}$$

where a_1 and a_2 constants that are defined by the initial conditions u_0 and u_1. Since $|\phi| > 1$, it follows that

$$u_n \approx a_1\phi^n \tag{7.20}$$

when n is large, so (7.18) holds true.

The next appearance of ϕ is somewhat more surprising. It arises in connection with Pascal's triangles, which were developed in the 17th century for the computation of binomial expansions. Pascal's idea was remarkably simple - he constructed a triangle of the form shown in Table. 7.3, in which each interior element is obtained by adding the two numbers directly above it (note, for example, that the closest numbers above 10 are 4 and 6). He further established that the numbers in each row represent coefficients in the binomial expansion. Thus, the terms in row 4 correspond to

$$(1 + x)^4 = x^4 + 4x^3 + 6x^2 + 4x + 1 \tag{7.21}$$

0					1				
1				1		1			
2			1		2		1		
3		1		3		3		1	
4	1		4		6		4		1
5	1	5		10		10		5	1
⋮					⋮				

Table 7.3. Pascal's triangle.

To see how Fibonacci numbers are connected with this result, it is useful to rearrange the triangle in the manner shown in Table 7.4. If we add the numbers along each diagonal, we obtain

$$\{1, 2, 3, 5, 8, 13, 21, 34, 55, 89, \ldots\} \tag{7.22}$$

which are precisely the elements of the Fibonacci sequence. This is truly unexpected (and aesthetically pleasing, if I may say so).

			3	5	8	13	21	34	55	89
0	1									
1	1	1								
2	1	2	1							
3	1	3	3	1						
4	1	4	6	4	1					
5	1	5	10	10	5	1				
6	1	6	15	20	15	6	1			
7	1	7	21	35	35	21	7	1		
8	1	8	28	56	70	56	28	8	1	
9	1	9	36	84	126	126	84	36	9	
10	1	10	45	120	210	...				
⋮				⋮						

Table 7.4. Rearranged Pascal's triangle.

The Fibonacci sequence and the number ϕ are also associated with various natural forms and phenomena. They are directly related to the geometric pattern of the sunflower and the nautilus shell, the mathematical description of bee colonies and even atomic physics.[47] The following example demonstrates the connection between ϕ and the logarithmic spiral, which is a commonly encountered form in nature.

Example 7.8. Many living organisms (such as the nautilus, for example), have shells whose form is a logarithmic spiral. To see how this geometric figure is related to ϕ, consider the rectangle shown in Fig. 7.8. It is readily observed from this figure that the radii of arcs BD and BW are l and l/ϕ, respectively.

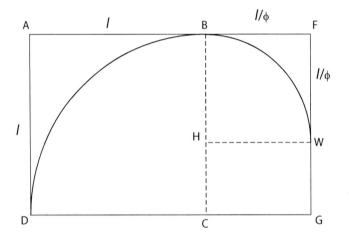

Fig. 7.8. The first step in constructing a logarithmic spiral.

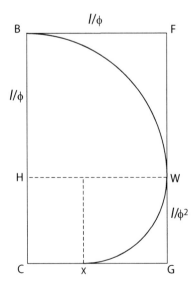

Fig. 7.9. The second step in the construction.

Let us now focus on rectangle $BFCG$, and partition it as shown in Fig. 7.9. Since ϕ satisfies the equation

$$\phi^2 - \phi - 1 = 0 \tag{7.23}$$

it follows that

$$\phi(\phi - 1) = 1 \tag{7.24}$$

and therefore

$$\phi - 1 = \frac{1}{\phi} \tag{7.25}$$

as well. This allows us to express WG in terms of ϕ as

$$WG = l - \frac{l}{\phi} = \frac{l(\phi - 1)}{\phi} = \frac{l}{\phi^2} \tag{7.26}$$

If we continue this process and connect all the arcs, the resulting figure will be a logarithmic spiral, whose length can be easily calculated as

$$L = \sum_{k=0}^{\infty} \frac{l\pi}{2} \cdot \frac{1}{\phi^k}. \tag{7.27}$$

Some Concluding Remarks

We will close this section with several brief remarks regarding the role that beauty plays in science and mathematics. The preceding discussion clearly indicates that aesthetic criteria can help us gain new insights into the nature of physical reality, and the laws that govern it. This is particularly evident in areas such as string theory or cosmology, where experimental results are difficult (or even impossible) to obtain. It would be a mistake to conclude, however, that aesthetic principles *always* lead to unanticipated results and major breakthroughs. More often than not they are simply a source of pleasure, which provides scientists and mathematicians with additional motivation for their work. This may not be a particularly "glamorous" function, but its importance should not be underestimated. Ultimately, this sense of aesthetic appreciation may be precisely what attracts scientists and mathematicians to their profession, even when they are not directly aware of it. If we agree that this is indeed the case, it would not be an exaggeration to claim that beauty represents one of the cornerstones of the scientific enterprise. Such a statement would certainly resonate well with theologians, who see beauty as a foundational principle that permeates all of reality.

7.4   Notes

1. http://en.wikiquote.org, 2005.

2. Quoted in: Robert Root-Bernstein, "The Sciences and Arts Share a Common Creative Aesthetic," in *Aesthetics and Science*, Alfred Tauber (Ed.), Kluwer Academic Publishers, 1997.

3. Quoted in: Stephen Clark, *Biology and Christian Ethics*, Cambridge University Press, 2000.

4. See, for example: C. Martindale, P. Locher and V. Petrov (Eds.), *Evolutionary and Neurocognitive Approaches to Aesthetics, Creativity and the Arts*, Baywood Publishing Company, 2007.

5. Early Christian thinkers such as Justin the Martyr (c. 100-160), Irineaus (c. 130-200) and Origen (c. 185-254) gave a prominent role to beauty in their writings.

6. Alejandro Garcia-Rivera, *The Community of the Beautiful*, Liturgical Press, 1999.

7. Clark, *Biology and Christian Ethics*.

8. S. Chandrasekhar, *Truth and Beauty: Aesthetics and Motivations in Science*, University of Chicago Press, 1987.

9. Richard Viladesau, *Theology and the Arts: Encountering God through Music, Art and Rhetoric*, Paulist Press, 2000.

10. Roger Scruton, *Beauty*, Oxford University Press, 2009.

11. Umberto Eco, *Art and Beauty in the Middle Ages*, Yale University Press, 2002.

12. See *e.g.* Carl Jung *et al.* (Eds.), *Archetypes and the Collective Unconscious*, Princeton University Press, 1980.

13. Carl Jung (edited by Aniela Jaffe), *Memories, Dreams, Reflections*, Random House, 1963.

14. Diarmuid O'Murchu, *Quantum Theology*, Crossroad, 2003.

15. H. E. Huntley, *The Divine Proportion*, Dover, 1970.

16. Garcia-Rivera, *The Community of the Beautiful*.

17. Quoted in: Arthur Koestler, *The Act of Creation*, Penguin Group, 1990.

18. Richard P. Feynman, *The Meaning of It All: Thoughts of a Citizen-Scientist*, Addison-Wesley, 1998.

19. http://www.worldofquotes.com, 2005.

20. Albert Einstein, *Ideas and Opinions*, Random House, 1988.

21. John Polkinghorne, *Belief in God in an Age of Science*, Yale University Press, 1998.

22. This point was recognized as early as the 14th century, by philosopher William of Occam. The famous "Occam's razor" claims that the simplest theory is always the best one (all other things being equal). Modern algorithmic information theory supports such a view, and maintains that we can explain a phenomenon *only* if the law is substantially simpler than the data.

23. Gottfried Wilhelm von Leibniz, *Discourse on Metaphysics and The Monadology*, Dover Publications, 2005.

24. Einstein, *Ideas and Opinions*.

25. Leonard Shlain, *Art and Physics*, Harper Collins, 1991.

26. Werner Heisenberg, *Physics and Philosophy*, Prometheus Books, 1999.

27. Quoted in: Shlain, *Art and Physics*.

28. Quoted in: Root-Bernstein, "The Sciences and Arts Share a Common Creative Aesthetic."

29. Aldous Huxley, "The Monocle," in *Collected Short Stories*, Chatto and Windus, 1969.

30. Quoted in: Paul Davies, *God and the New Physics*, Simon and Schuster, 1983.

31. http://en.wikiquote.org, 2005.

32. John Keats, "Ode on a Grecian Urn," in *Lyric Poems*, Dover Publications, 1991.

33. Paul Krugman, "How Did Economists Get It So Wrong?", *New York Times Magazine*, September 6, 2009.

34. Koestler, *The Act of Creation*.

35. The decomposition algorithm used to obtain the structure in Fig. 7.5. was proposed in: A. I. Zečević and D. D. Šiljak, "Balanced decompositions of sparse systems for multilevel parallel processing," *IEEE Transactions on Circuits and Systems*, **CAS-41**, pp. 220-233, 1994.

36. Root-Bernstein, "The Sciences and Arts Share a Common Creative Aesthetic."

37. E. G. Mazurs, *Graphic Representations of the Periodic System During One Hundred Years*, University of Alabama Press, 1974.

38. Root-Bernstein, "The Sciences and Arts Share a Common Creative Aesthetic."

39. Ibid.

40. Antoine de Saint-Exupery, *The Little Prince*, Harcourt, 2000.

41. A modern application of the coloring problems is the assignment of broadcasting frequencies to cell phone transmitters. In this case, the objective is to ensure that no two neighboring transmitters have the same frequency (since that would cause interference).

42. Philip Davis and Reuben Hersh, *The Mathematical Experience*, Birkhauser, 1981.

43. Quoted in: Gary Flake, *The Computational Beauty of Nature*, MIT Press, 2000.

44. Plato, *The Republic*, Penguin Group, 2003.

45. This connection was pointed out by Davis and Hersh, in *The Mathematical Experience*.

46. Equation (7.13) is derived from the Pythagorean theorem

$$AD^2 = l^2 + \left(\frac{l}{2}\right)^2 \tag{7.28}$$

and the fact that

$$AD = AE + \frac{l}{2} \tag{7.29}$$

Solving for AE and observing that $AC = AE$, (7.13) follows directly.

47. Huntley's book *The Divine Proportion* provides several examples of this. See also: Mario Livio, *The Golden Ratio*, Broadway Books, 2002.

Chapter 8

Ethics, Science and Theology

"If geometry were as much opposed to our passions and interests as is ethics, we would contest it and violate it but little less, not withstanding the demonstrations of Euclid." *Gottfried von Leibniz* [1]

"Even philosophers who write books in praise of humility take care to put their names on the title page." *Cicero* [2]

There is general agreement among both scientists and theologians that ethics plays a central role in their respective fields of inquiry. It should come as no surprise, however, that these two disciplines strongly disagree on a number of important issues that arise in this context. In the following sections we will focus our attention on two questions that have been particularly controversial – the existence of free will, and the principle of moral relativism. These questions expose some deep discrepancies between the secular and religious views of morality, and suggest that their positions may be impossible to reconcile. We will argue, however, that this may be an overly pessimistic conclusion, and that the pursuit of an "integrated" approach to ethics is not an unrealistic goal. As in the case of aesthetics, we will find that science and theology share a common set of values, and that these similarities can sometimes overshadow the differences that separate them.

Before we proceed to examine some of these questions in greater detail, it might be helpful to make several preliminary observations regarding the methods and objectives of ethical inquiry. Perhaps the most important one is that ethical decisions entail human choices, which are inherently unpredictable. As a result, it is fair to say that ethicists are usually in no position to anticipate how an individual will act in a given situation. What they *can* do, however, is investigate how one *ought* to act under such circumstances. This type of analysis usually requires the formulation of general rules and guidelines for "proper" behavior, as well as precisely defined criteria for what is "right" and what is "wrong." It is in this domain that ethical outlooks tend to collide, and branch out into different (and often incompatible) theories.

In this chapter we will consider three distinct positions regarding the nature and function of ethics. The first two approaches (which are loosely described as "evolutionary" and "relativistic" ethics) are purely secular, while the last one represents a theological view of morality. A brief overview of their main features is provided below.

Evolutionary Ethics. The view referred to as "evolutionary" ethics has been promoted by a number of prominent biologists and sociobiologists. Its principal claim is that our moral code is a consequence of evolution, and is merely an adaptation to collective living. These thinkers deny the possibility of free will, and maintain that all mental activity is determined by the underlying biochemical processes. From their perspective, ethics is no more than a collection of useful social rules, whose primary function is the survival of the species.

Relativistic Ethics. The so-called "relativistic" approach concedes that ethics is a legitimate field of inquiry, but denies that it has an absolute character. Thinkers of this persuasion believe in the multiplicity of moral truths and the relative nature of their claims. From this standpoint, what matters is the identification of a desirable objective. Once that is done, the proper course of action can be determined by logic and scientific reasoning.

Religious Ethics. The position of "religious" ethics is that there is such a thing as absolute moral truth. According to this view, in any given situation we do have an intrinsic sense of what is good and moral (although we don't always follow it). Such claims are rooted in the belief that human conscience is not just a product of social norms, but is also a reflection of a "higher" truth that represents the ground of our being.

The three paradigms outlined above are fundamentally different, and their treatment of ethical problems often leads to conflicting solutions. The literature devoted to these controversies is very extensive, and clearly cannot be summarized in a book of this scope. Our approach in the following will be far more modest, and will focus on several specific issues where analogies from the world of science may help illuminate the problem.

8.1 Free Will and Determinism

New insights from recent research in the fields of biology and related disciplines have given rise to a wave of publications regarding the origins of religion and morality. A prominent theme in these writings has been the suggestion that religion exists not because there is such a thing as a "divine reality", but because religious feelings are good for survival. Thinkers who subscribe to this point of view tend to be thoroughly reductionist, and flatly deny the existence of free will:

> "Modern science directly implies that there are no inherent moral or ethical laws. ... Free will, as traditionally conceived, the freedom to

make uncoerced and unpredictable choices among alternative possible courses of action, simply does not exist. The evolutionary process cannot produce a being that is truly free to make choices." *William Provine* [3]

If correct, a view like Provine's would undermine the very foundations of ethics, since the ability to make free choices is a necessary condition for moral responsibility. Indeed, if our desires and motives were fully determined by chemical processes in the body, we could not be held accountable for our actions. Whatever we do would simply follow from the laws of physics, and morality as we know it would collapse.

In order to examine the validity of such claims, we first need to consider what the term "fully determined" means to scientists. When we say that a physical process is *deterministic*, we normally assume that it is governed by some sort of natural law (or set of laws) that can be precisely described in mathematical terms. Scientists have formulated many such models over the past four centuries, and have successfully used them to predict the behavior of a wide variety of physical systems. In the case of highly complex systems, however, we can legitimately question whether such a description is possible. To understand the source of the problem, we should first recall that science normally deals with *isolated* subsystems, whose complexity is made manageable by disregarding a broad range of external influences. There are many cases where such approximations work very well, and can be applied without significant loss of accuracy (electric circuits are a typical example). On the other hand, there are also phenomena such as chaos or the EPR effect in quantum mechanics, where even the smallest details matter. Systems of this type must be treated as *indivisible wholes*, since any simplification in their mathematical description necessarily leads to completely unreliable results. This property severely restricts our ability to make meaningful predictions, and precludes us from establishing causal relationships of any kind.

The inadequacy of mathematical models can also be analyzed from the standpoint of algorithmic information theory, which maintains that our entire concept of scientific knowledge assumes "compression" of data into compact laws. Such a compression is possible whenever the data conform to a pattern, in which case we can provide an "explanation" for the observed phenomenon. Modern science has shown, however, that there are numerous instances where this kind of reduction is not possible, and the law is just as complicated as the data itself. In such cases mathematical models are meaningless, and provide no useful information about the system.

The preceding analysis points to some fundamental limitations of physical modeling, and suggests that complex systems do not always lend themselves to precise analytic descriptions. In many such cases, we actually have no way of telling whether the system behavior is random or lawful. If the mental processes that underlie our thoughts and actions are anything like this, it is reasonable to assume that we will never be able to properly characterize them in mathematical terms. Without such a description, it is hard to see how one could ascertain whether human decisions are fully determined or not. The best we could do is speculate.

It is questionable whether this kind of uncertainty would be eliminated even if we somehow managed to formulate a precise mathematical model of the human brain. To see why this is so, let us consider a simple thought experiment in which we will assume that our mental processes can be completely described by a system of differential equations. We will additionally assume that the initial conditions for this system are well defined, and can be measured with a high degree of accuracy. Since such a model contains no uncertainties, it would be reasonable to assume that all human thoughts and actions could be predicted with the help of a sufficiently powerful computer. It would then be entirely appropriate to say that our decisions are "fully determined" by the underlying equations and initial conditions.

It turns out, however, that our current understanding of nonlinear system theory points to a rather different conclusion. An illustration from chaos theory might help explain the logic behind this seemingly paradoxical claim. As is well known, the equations that describe chaotic dynamics are completely deterministic, but their solutions nevertheless display an apparent randomness and extraordinary sensitivity to external influences. The extent of this sensitivity is illustrated in Fig. 8.1, where we compare two trajectories whose initial values differ by 0.000001 % (for better clarity, the trajectories are represented by lines of different thickness). It is readily observed that the two solutions are identical at first, but bear no resemblance to each other after a sufficiently long time.

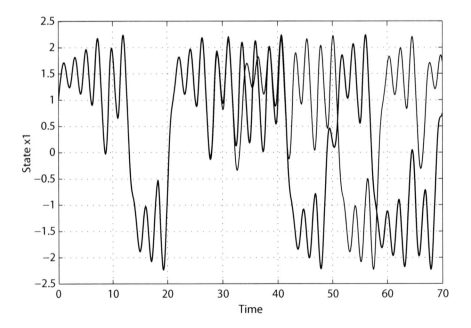

Fig. 8.1. Evolution of two initially close trajectories.

The above example indicates that even if we allow for the existence of a perfectly accurate physical model of the human brain, the specific outcomes may be so sensitive to microscopic changes that they are unpredictable for all practical purposes. Eliminating this uncertainty would require *infinite precision*, which is something that no computer is capable of doing. Such a possibility clearly undermines the deterministic hypothesis, and suggests that we may not be able to resolve the free will problem by purely scientific considerations.

One could argue, of course, that the above discussion doesn't rule out a scenario in which our thought processes would be fully controlled by a deterministic law that is simply too complex to be identified. If we were to adopt such an outlook, we could still claim that the notion of free will is just an illusion, which is a product of our epistemological and technological limitations. With that in mind, it would be interesting to see if we can identify a class of relatively simple deterministic natural laws that allows for some form of *irreducible* uncertainty, even in cases when our knowledge of the system parameters is assumed to be *infinitely* precise.

Schrödinger's equation in quantum mechanics provides us with a classic example of such a model. This equation is completely deterministic, but its solutions allow us to compute only the *probabilities* of different outcomes. According to modern physics, no amount of additional information can remove this inherent uncertainty. In order to see how this line of reasoning might be extended to the free will problem, it is helpful to recall that an undisturbed particle is very rarely found in a definite state which can be described precisely. Instead, the state of such a particle normally consists of many coexisting alternatives that are mutually exclusive in the world that we are familiar with. However, the moment we choose to perform a measurement, the particle collapses into a single state, thus avoiding any contradiction with our everyday experience. We cannot tell which state will emerge from this process – the best we can do is provide the probabilities for different outcomes (see Chapter 4 for more details).

Based on the properties described above, let us now consider a second thought experiment, in which we will assume that human consciousness allows for the simultaneous existence of mutually exclusive choices (much like a quantum system).[4] Taking this analogy a step further, we could also assume that a probability can be assigned to each outcome of the decision making process, based on some deterministic mathematical model. This, however, would be all the information that this model can provide. Under such conditions, we could reasonably claim that our behavior is governed by some overarching physical laws and constraints, while maintaining that our choices are unpredictable in an absolute sense. As in quantum mechanics, the probabilistic nature of the model would not be a result of our ignorance of certain relevant parameters in the system. Instead, it would reflect an irreducible uncertainty that is intrinsic to the operation of the human mind.

What conclusions can we draw from all this? It seems to me that what we know about complex systems (both of the macroscopic and microscopic variety) clearly suggests that freedom of choice is at the very least a rational possibility. Modern science recognizes that many of nature's basic processes are inherently uncertain, and there is no a priori reason to believe that the workings of our mind should be an exception in that respect. When it comes to human decisions, I am inclined to agree with Mark Twain's witty observation that:

"Prediction is difficult, especially of the future." [5]

8.2 Religious Ethics and Moral Relativism

Throughout the 20th century many philosophers have argued that ethical knowledge is relative. According to this view, there are no absolute "musts" that are divorced from a purpose. There may be many possible goals which meet the standards of rationality, but once a choice has been made, facts and reason alone should determine the proper course of action.

> "I do not think there is, strictly speaking, such a thing as ethical knowledge. ... Given an end to be achieved, it is a question for science to discover how to achieve it. All moral rules must be tested by examining whether they tend to realize ends that we desire. I say ends that we desire, not ends that we "ought" to desire. What we "ought" to desire is merely what someone else wishes us to desire."
> *Bertrand Russell* [6]

The view that multiple goals are necessary is certainly consistent with historical experience. Humans have never been able to agree on a single ethical standard, and what constitutes decent behavior varies greatly across civilizations. With that in mind, it seems perfectly natural to assume that "absolute" goals and criteria simply don't exist, and that moral relativism is the only viable option.

The position of religious ethics in these matters is a very different one. According to St. Thomas Aquinas, the ultimate purpose of morality is the contemplation of God, and the proper way to achieve this is through *agape*, a self-transcending love which does not require satisfaction of any kind. What distinguishes this attitude from the one proposed by moral relativists is the fact that Christianity does not offer strictly defined objectives, but rather a *way of life*. It is assumed that if one leads a life guided by agape, moral decisions will come naturally, and will adapt to the time, place and social circumstances. Those who follow this path will gradually approach the goal without ever having to explicitly specify its characteristics. St. Augustine elegantly articulated this outlook in the simple phrase:

"Love God, and do what you please." [7]

A natural starting point for comparing these two ethical paradigms is to examine whether the notion of "absolute truth" is rationally acceptable. From a scientific standpoint, an interesting way to approach this question would be to draw an analogy with special relativity. According to Einstein's theory, different observers will necessarily record different times and locations for the same event.

What they actually see depends on the relative speed with which they move, and there is no "privileged" viewpoint that provides a "correct" description of the event in question. In that respect, nature seems to be very democratic.

If we were to end our discussion at this point, it would appear that Einstein's theory is consistent the views of moral relativists. If what is "right" truly varies from observer to observer, then it makes sense to assume that no interpretation of morality is absolute. And yet, there is a subtle twist to this argument that undermines such a conclusion. To see this more clearly, we need to consider how the world would appear to an individual traveling at the speed of light. This is the extreme case of special relativity, where time effectively "stops," and all events become simultaneous. The viewpoint of such an observer would be "absolute" in a very real way, since for her all temporal differences become irrelevant. The past and future would merge into an "eternal" present moment, and her conventional notion of causality would become obsolete (since it makes no sense to say that one event regularly precedes another one). It is important to keep in mind, however, that such a frame of reference is not accessible to us. This is because the mass of any object tends to infinity as it approaches the speed of light (see Chapter 5 for more details). Consequently, we may conclude that an "absolute" system of reference exists in theory, but cannot be experienced by human beings due to their material nature.

The proposed analogy with special relativity suggests that what we perceive as "irreconcilable differences" in ethical standards may really be a result of our inherent limitations, and our inability to grasp reality as a whole. From that perspective, it would be entirely plausible to argue that there *is* a universal moral outlook in which seemingly disparate views become unified on a higher level. The fact that we cannot articulate these "higher truths" with any precision doesn't automatically imply that they do not exist. It may instead be viewed as an indication that certain forms of knowledge are far too complex to be fully understood.

> "When two texts, or two assertions, perhaps two ideas, are in contradiction, be ready to reconcile them rather than cancel one by the other; regard them as two different facets, or two successive stages, of the same reality, a reality convincingly human just because it is complex." *Marguerite Yourcenar* [8]

Even if we agree that there is such a thing as "absolute moral truth," it remains unclear how this kind of knowledge might be acquired. Virtually all religious traditions claim that this entails some form of spiritual development, but their specific teachings on the subject tend to be very different (and often contradictory). In view of that, one could possibly formulate a somewhat "weaker" version of moral relativism, which maintains that this approach to ethics is the only *practical* option (although theory allows for other possibilities as well). Such a view appears to have some merit, particularly if one considers how easily different visions of the "ultimate good" can turn into armed conflicts. There is ample evidence of this throughout recorded history, which provides us with an endless list of religious and ideological wars.

In examining this question, it is important to keep in mind that all monotheistic traditions implicitly agree that the contemplation of God is the "ultimate good," and the final goal of human existence. Although they propose rather different paths to this goal, they do share the belief that it cannot be reached without cultivating some form of *selfless love* toward other human beings. This theme is particularly prominent in Christian theology, which sees *agape* as the only way to overcome the cultural, historical and economic barriers that separate us.

> "Love alone can transform itself according to the concrete demands
> of every individual and social situation without losing its eternity,
> dignity and unconditional validity." *Paul Tillich* [9]

The notion that love can overcome our differences and somehow unify all of humanity is certainly an appealing one. We must concede, however, that living a life guided by *agape* is a difficult thing to do. Many who see themselves as "religious" have taken a rather different approach, and have used their beliefs to justify intolerance (and even violence) toward those who do not share their views. Although this is (and always has been) a widespread phenomenon, I would be reluctant to associate it with the true essence of religion. Nor would I describe individuals who behave in such a way as "genuine" believers. I would argue instead that no faith can be divorced from its fundamental ethical principles, and that belonging to a particular religious tradition cannot be reduced to the mere observance of its rituals. True spiritual development requires much more, and entails a great deal of personal responsibility. With that in mind, it may perhaps be more accurate to describe a "genuine believer" as an individual who embraces love, faith and humility as central virtues, and shapes his or her life accordingly.

Can narrowing the notion of a "believer" in the manner outlined above help us bridge the gap that separates the moral teachings of different religious traditions? Perhaps it can. If all monotheistic belief systems truly agree that the contemplation of God is the ultimate spiritual goal, then this places certain limits on what can constitute acceptable ethical behavior for their "genuine" adherents. Indeed, despite their other differences, I am not aware that any mainstream religion condones abject poverty, a complete disregard for human life or extreme cruelty. These universal moral constraints are presumably what Dostoevsky had in mind when he wrote:

> "If God doesn't exist, everything is permissible." [10]

An analogy with chaos theory provides a nice illustration for this argument. To understand the connection, we should first recall that when a nonlinear system is in a chaotic regime, even the most miniscule variations in initial conditions produce trajectories that become completely different after a sufficiently long time. As a result of this sensitivity, there is tremendous diversity among the solutions, even if they all start out very close to each other. It turns out, however, that this apparent diversity is constrained by the form of the common final 'goal' (*i.e.* the attractor). In Fig. 8.2 we show an example of such an object, which acts like a sort of "magnet" that draws different trajectories to itself. This figure also shows two different solutions whose initial points are labeled by dark circles (for better clarity, one of them is represented by a solid line and the other by a dashed one).

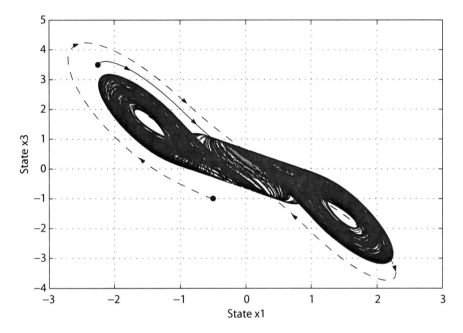

Fig. 8.2. Projection of the attractor onto the $x_1 - x_3$ plane.

In a subtle way, the existence of an attractor imposes certain restrictions on the kind of dynamic behavior that is possible. Indeed, although the system trajectories can traverse this strange geometric object in many different ways, the "distances" between them cannot exceed the dimensions of the attractor. With that in mind, it is not unreasonable to speculate that constraints of a similar nature might apply to religious believers as well. It could be argued, for instance, that despite our individual and cultural differences, the actions of those who are *genuinely* drawn to a common spiritual goal cannot deviate too much from each other. In the absence of such an "attractor," however, there would be no limits to human actions (as Dostoevsky rightly observed). In a world where moral relativism prevails, every kind of behavior is potentially permissible, and ethics can easily degenerate into a tool for justifying the actions of the powerful. The 18th century Prussian king Frederick the Great articulated this view with the candor of a true absolute monarch:

> "I begin by taking. I shall find scholars later to demonstrate my perfect right." *Frederick the Great* [11]

If we allow that positing a single spiritual goal is a rationally justifiable possibility (which can potentially lead to a unified moral outlook), we should also consider whether such a choice has any *practical* advantages. In thinking about this issue, it is useful to observe that the existence of multiple goals can give rise to some serious ethical conflicts. To see why this is so, consider equality and freedom, both of which are seen as desirable and fully rational goals in contemporary society. It is obvious that these two objectives cannot be attained simultaneously, since total liberty for the powerful is not compatible with a decent existence for the weak. The same can be said about individual rights and national security, which become almost impossible to reconcile in times of war or political crises.

How should one decide between such conflicting goals in a relativistic moral system? There don't seem to be any clear criteria, since the fact that values clash does not automatically imply that one is right and the other is not. Indeed, it is by no means obvious that rigorous law enforcement is any better than compassion, or that liberty supersedes equality (the French republic actually gives them equal weight, as witnessed by its motto – Liberté, Egalité, Fraternité). In light of these apparent ambiguities, the best that 'relativist' morality can offer is a kind of fragile equilibrium, which in some sense "minimizes the damage." This may be rather modest, but is clearly better than the alternative, since accepting one of the competing options as the exclusive objective potentially exposes society to all kinds of excesses (communism and unconstrained capitalism come to mind immediately).

In contrast to the relativist paradigm, the goal of Christian ethics is both unique and transcendental. No specific set of actions (or rules) is automatically conducive to the contemplation of God - this requires a complete change of mindset and attitude to life. As a result, in the context of Christian morality it may be entirely consistent to favor liberty in one situation and equality in another. Both of these goals are simply means, and circumstances will determine which one is appropriate at a given time. Such a framework provides a great deal of practical flexibility, since there is no need to develop specific criteria for 'right' actions. The medieval mystic Meister Eckhart held that such rules can actually distract us from the true purpose of morality:

> "People should think less about what they ought to do and more about what they ought to be." *Meister Eckhart* [12]

8.3   The Role of Values in Science

It has often been said that science is a purely objective enterprise, which has little use for anything but reason. In this field where logic reigns supreme, traditional religious values appear to have little significance, as does the kind of subjective judgement that is normally associated with the arts and humanities. Our objective in the following will be to examine whether this is indeed the case, and whether it makes sense to speak of science as a completely "autonomous" discipline. In addressing these questions, we will take into account "external" factors such as theological virtues, aesthetic preferences and sociopolitical circumstances, all of which influence the way scientific decisions are made. At the end of this section, we will also take a closer look at certain psychological mechanisms that affect the way we conduct research and interpret our data.

Science and Theological Virtues

We begin with a brief analysis of the role that theological virtues such as *humility* and *faith* play in the domain of science. Most practicing scientists will agree that a certain amount of humility is essential to their work, since researchers (just like all other individuals) need to embrace the wisdom of those who have been successful in the field. This obviously requires the recognition that some people in our profession know more than we do, and can teach us a thing or two. More importantly, humility is also instrumental in reminding scientists that their knowledge is inherently limited. The consequences of ignoring this fact can be very serious indeed:

> "Man, formerly too humble, begins to think of himself as almost God. ... In all this, I feel a grave danger, the danger of what might be called cosmic impiety. The concept of "truth" as something outside human control has been one of the ways in which philosophy hitherto has inculcated the necessary element of humility. When this check upon pride is removed, a further step is taken on the road towards a certain kind of madness. ... I am persuaded that this intoxication is the greatest danger of our time." *Bertrand Russell* [13]

The "intoxication" that Russell refers to is perhaps most apparent in the various abuses of scientific knowledge (our treatment of the environment is a glaring example). From that standpoint, it could be legitimately argued that certain projects should never be pursued, despite the fact that they may be technically feasible. When it comes to such decisions, scientists would do well to adopt the perspective offered by Isaac Newton:

> "I do not know what I may appear to the world; but to myself I seem to have been only like a boy playing on the seashore, and diverting myself in now and then finding a smoother pebble or prettier shell than ordinary, while the great ocean of truth lay all undiscovered before me." [14]

In the context of science, humility also implies an openness to novelty and unexpected ideas. When seen in that light, the Christian teaching that "we should be like children" actually contains some useful advice for researchers, since children tend to have fewer prejudices and are often in a better position to experience some of the subtler manifestations of reality. Physicist Robert Oppenheimer (who is sometimes referred to as the "father" of the atomic bomb) openly acknowledged this possibility:

> "There are children playing in the street who could solve some of my top problems in physics, because they have modes of sensory perception that I lost long ago." [15]

Along these lines, it is interesting to note that as a boy Einstein repeatedly wondered what the world would look like if viewed from a beam of light. There is no doubt that such a question would have been dismissed as absurd in the academic circles of the time. And yet, Einstein's youthful openness and imagination ultimately lead to one of the greatest breakthroughs in the history of science.

Another traditional religious virtue that plays a significant role in science is *faith*. This may perhaps sound strange, given the secular orientation of many scientists and the central role of empirical reasoning in their work. Nevertheless, recent developments in the philosophy of science suggest that this virtue is an essential prerequisite for creativity. Indeed, serious scientific investigation generally requires faith in the ideas and assumptions that guide the work. If continuous progress were a necessary condition for research, many successful scientific enterprises would have been abandoned early on, due to temporary difficulties and discrepancies. What sustains our continuous efforts despite such adversity is precisely the belief that we are on the right track. It is true, of course, that this is not exactly the same kind of belief that we encounter in religion. But there is enough similarity between the two to warrant the assumption that they may share the same origins.

In thinking about this subject, we must also consider what happens if a belief ultimately turns out to be wrong. Would this mean that our efforts have been completely wasted? Philosopher Thomas Kuhn addressed this issue in his classic book *The Structure of Scientific Revolutions.*[16] He made the point that an unquestioned belief in a paradigm can be a *good* thing for science, even if the paradigm is imperfect. If we were to constantly evaluate our basic principles, or attempt to replace them as soon as the first failure is encountered, not much would be accomplished. Scientific paradigms do change, of course, but this does not happen easily.

> "In this way, a paradigm is like a well-shielded and well-designed bomb. A bomb is supposed to blow up; that is its function. But a bomb is not to supposed to blow up at any old time; it's supposed to blow up in very specific circumstances. A well designed bomb will be shielded from minor buffets. Only a very specific stimulus will trigger the explosion." *Peter Godfrey-Smith*[17]

The trigger for this kind of change is often a paradox, whose emergence heightens the contrast between reality and the prevailing scientific views. Something like this happened at the beginning of the 20th century, when the inability of science to explain certain experimental data brought about a fundamental change in our understanding of matter, space and time. This process (which was by no means painless) obviously had a positive effect, since it opened the door to new ideas that ultimately lead to the development of quantum mechanics and the theory of relativity.

The preceding discussion strongly suggests that human values play an important role in defining both research priorities and the standards that guide the investigation. In view of that, it would be perfectly sensible to claim that the pursuit of scientific knowledge requires a certain kind of *moral preconditioning.* It is important to add, however, that the scientific method is *not* the proper framework for investigating the nature and origins of this preconditioning. This type of inquiry is usually the prerogative of philosophy and theology.

So what do philosophers and theologians have to say on the subject? While they may not see eye to eye on many issues related to ethics, they will probably agree that virtues such as humility, perseverance and honesty cannot be developed before one comes to *believe* that these qualities are desirable. Most of them would also concede that such beliefs are usually adopted from the tradition in which we are brought up. Given that traditional values often find their roots in religion, theologians could take this argument a step further, and claim that certain elements of religious morality must be in place *before* scientific reasoning can be properly utilized. When viewed from that perspective, the connection between science and religion would appear to be deeper than most of us suspect.

The Influence of Aesthetic and Social Factors

In addition to traditional virtues, there are a number of other "external" factors that influence research practices in mathematics and the natural sciences. As an illustration of this fact, consider the way in which mathematicians determine what problems and research directions are sufficiently relevant to be pursued. Given that some 200,000 new theorems are published each year in professional journals,[18] what possible "scientific" criteria could we have for establishing their relative importance?

> "If the number of theorems is larger than one can possibly survey, who can be trusted to judge what is 'important'? One cannot have survival of the fittest if there is no interaction. ... Because of this, the judgment of value in mathematical research is becoming more and more difficult, and most of us are becoming mainly technicians."
> *Stanislaw Ulam* [19]

The problem that Ulam refers to is further exacerbated by the fact that mathematical results cannot be compared against some hypothetical "mathematical reality." Unlike the physical sciences, where theories are evaluated by how well they conform to experimental observations, mathematics requires nothing more than internal consistency. Under such circumstances, one could legitimately argue that the best criteria for judging relevance are often *aesthetic* ones. Carl Friedrich Gauss hinted at this possibility some 200 years ago, when he argued that the beauty of a mathematical proof is just as relevant as its novelty.

> "Sometimes one does not at first come upon the most beautiful and simplest proof. ... The finding of new proofs for known truths is often at least as important as the discovery itself." [20]

In his famous essay *A Mathematician's Apology*, Cambridge mathematician G. H. Hardy made a similar claim:

"A mathematician, like a painter or poet, is a maker of patterns. The mathematician's patterns, like the painter's or poet's, must be beautiful; ideas like the colors or words must fit together in a harmonious way. Beauty is the first test: there is not a permanent place in the world for ugly mathematics." [21]

The practice of using aesthetic criteria in making research-related decisions is by no means unique to mathematics. Scientists make such choices on a regular basis, and frequently elect to promote certain research directions at the expense of others based on aesthetic preferences. A striking example of this kind is Hermann Weyl's gauge theory of gravitation, which sought to unify gravity and electromagnetism under a common mathematical framework. Although Weyl came to believe that his theory was physically unrealistic, he held on to it simply for the sake of its beauty. [22] Interestingly, many years later it turned out that Weyl's intuition was correct, and his principle of gauge invariance was incorporated into quantum field theory.

Examples such as the one provided above suggest that a developed sense of beauty can be an important element in the process of scientific discovery. With that in mind, it would not be inappropriate to say that the successful practice of science requires a certain amount of *aesthetic preconditioning*. As in the case of ethics, much of this preconditioning takes place at a very young age, and is closely associated with the tradition in which we were raised.

In evaluating how scientific decisions are made, we must also take into account the prevailing social and political circumstances. During the Cold War, for example, the United States and the Soviet Union allocated enormous resources for aerospace research, for reasons that had far less to do with scientific curiosity than with politics. This strategy clearly favored the development of areas such as control theory, electronics and communications, perhaps at the expense of theoretical physics (which was a research priority in the years leading up to the development of the atomic bomb).

Scientific research priorities are influenced by economic interests as well. The following excerpt from a Newsweek article provides a striking illustration of this point.

"Drug companies have largely given up on looking for [remedies for malaria]. From a commercial perspective, it makes little sense to turn out costly pharmaceuticals for people who can't afford shoes. ... Altruism has never played a big role in malarial research. Quinine enabled Europe to colonize the tropics. Chloroquine grew out of efforts to protect U.S. troops abroad. Without an empire or an army on the line, the developed world will need a new rationale for fighting malaria. At the moment, the best one is that 2.1 billion people – about 40 percent of the world's population – are in danger." [23]

The observations made above underscore the fact that research goals are not determined by scientists alone. In many cases, decisions of this sort are also shaped by individuals who control the sources of funding. As a result, it is not unusual to encounter situations where governments and private institutions promote potentially dangerous projects, which are beneficial only to certain interest groups.

How does the possibility of such abuses influence the way scientists approach their work? Perhaps the most obvious implication is that there are times when knowledge should not be pursued for its own sake. When it comes to projects that pose a potential danger to society or raise serious ethical questions, scientists cannot afford to focus exclusively on the technical aspects of the problem. They must also consult their *conscience*, since the choices they make can have serious consequences. Recognizing this fact and making the appropriate decisions clearly requires the kind of moral preconditioning that was discussed earlier in this section.

The Influence of Psychological Factors

Having seen how ethical, aesthetic and sociopolitical factors affect scientific research, it might be interesting to conclude this section with a brief examination of the role that human psychology plays in this process. We begin with the observation that the public perception of science and its value system often tends to be idealized.

> "The history of science is as inspiring in its human values as the legends of the saints." *W. S. Knickerbocker* [24]

> "A scientist is sometimes subjected to humiliation as his findings shift and invalidate some conclusion to which he has previously committed himself; but his loyalty to truth is such that he would rather cut off his right arm than suppress the new data." *P. B. Diedrich* [25]

But are scientists really paragons of intellectual honesty who will follow the inquiry wherever it may lead, regardless of whether this fits into their original plans and expectations? And will they always resist the temptation to interpret ambiguous data in a way that suits their interests? Modern research in psychology and sociology points to a rather different conclusion, and suggests that science has a reward structure which shapes the way research is conducted.

> "The organization of science consists of an exchange of social recognition for information. But, as in all gift-giving, the expectation of return gifts cannot be publicly acknowledged as the motive for making the gift." *Warren Hagstrom* [26]

In support of this argument, it is worth recalling that even some of the greatest names in the history of science were extremely keen on getting recognition for their work. Sir Isaac Newton, for example, personally appointed a committee of the Royal Society that would decide on the dispute between himself and Leibniz regarding the invention of calculus. Not surprisingly, the decision was in his favor (although Leibniz had published his work much earlier). Galileo, on the other hand, was in the habit of scrambling his correspondence in order to protect the secrecy of the experimental data that he gathered. It is interesting to note that his legendary conflict with the Catholic church was not just about heresy. It also involved Galileo's impatience to publish (and get credit for) a thesis which lacked adequate scientific proof. In doing so, he violated an agreement that he had previously reached with ecclesiastic authorities.

> "During his 1616 visit, Galileo received the support of some power-ful liberal theologians, particularly cardinals Roberto Bellarmine and Maffeo Berberini, who argued that if Copernicus' system was some-day proved true, then the church would have to reinterpret those biblical passages that seemed to contradict it. However, they also supported the compromise that Galileo eventually agreed to: Until such definitive proof was forthcoming, Galileo should discuss helio-centrism only hypothetically, and not promote it as a true description of the heavens. ... Convinced that he had the required proof in hand, Galileo published his *Dialogue on the Two Chief World Systems* in 1632. ... However, there was one problem: Galileo's new proof made no sense." *Timothy Moy* [27]

Of course, not all scientists have been so eager to gain the recognition of their peers and society. Many have demonstrated selfless dedication to their ideas, and have shown exemplary humility in the process. Nevertheless, historical evidence clearly shows that in most cases science is not just about discovery and sharing information, but also about who gets the credit. This is (and always has been) a powerful source of motivation, without which many discoveries would never have been made.

Another important psychological factor that affects scientific research has to do with the gathering and interpretation of experimental data. Psychologist Michael Mahoney [28] has argued quite persuasively that data are usually used to *defend* an existing belief, rather than to form a new one. Human beings seem to like their paradigms, and tend to change them only under extreme pressure. This is driven to some extent by our psychological makeup, which requires that we "filter" information prior to processing it (as a consequence of the fact that our brain can handle only a relatively small subset of the external stimuli that we receive). It is important to recognize that this filter is *not* unbiased, and often shapes our perceptions in a way that fits existing frameworks (the technical term for this effect is "constructive perception").

The obvious suggestion here is that what scientists actually perceive is nec-essarily conditioned by preexisting ideas, theories and expectations. Because of this subjective bias, important data may be overlooked. The problem is com-pounded by the fact that our methods of investigation place certain implicit restrictions on the conclusions that we can draw from empirical observations.

> "Astronomer Arthur Eddington once told a delightful parable about a man studying deep-sea life using a net with a three-inch mesh. After bringing up repeated samples, the man concluded that there are no deep-sea fish smaller than three inches in length. Our methods of fishing, Eddington suggests, determine what we can catch. If science is selective, it cannot claim that its picture of reality is complete." *Ian Barbour* [29]

It would appear, then, that scientists are just as fallible as other humans. Psychological factors such as the desire for recognition and stubborn resistance to new paradigms are integral parts of the profession, as are various types of value judgments. It is true, of course, that science possesses an admirable moral code, but it is also true that many of its central virtues are rooted in traditional morality. With that in mind, I would be very reluctant to say that there is an autonomous set of objective values and criteria that are unique to the scientific enterprise. Science may indeed be one of the most rational human activities, but this does not mean that it can be separated from the basic values of society.

> "One of the essential tenets of scientific rationality is the deep conviction that nothing should be accepted without sufficient proof or argument. ... However, no proof or argument can be given that one should be rational. If rationality is a value, then the decision to be rational is a moral choice." *Michael Heller* [30]

8.4   Notes

1. Quoted in: John Polkinghorne, *The Faith of a Physicist*, Fortress Press, 1996.

2. Quoted in: Will Durant, *The Story of Philosophy*, Simon and Schuster, 1972.

3. William Provine, "Progress in Evolution and Meaning in Life," in *Evolutionary Progress*, M. H. Nitecki (Ed.), University of Chicago Press, 1988.

4. The idea that the mind operates like a quantum system is not far-fetched, and has actually been seriously discussed by a number of authors. For a good introductory presentation of this topic, see:

   a) Evan Walker, *The Physics of Consciousness*, Basic Books, 2000.

   b) Fred Wolf, *Mind Into Matter*, Moment Point Press, 2001.

5. Quoted in: Gary Flake, *The Computational Beauty of Nature*, MIT Press, 2000.

6. Bertrand Russell, *Why I Am Not a Christian and Other Essays on Religion and Related Subjects*, Simon and Schuster, 1976.

7. Quoted in: Aldous Huxley, *The Perennial Philosophy*, Harper and Row, 1972.

8. Quoted in: Flake, *The Computational Beauty of Nature*.

9. Paul Tillich, *Morality and Beyond*, Harper Collins, 1981.

10. Fyodor Dostoevsky, *The Brothers Karamazov*, Random House, 1992.

11. Quoted in: Will and Ariel Durant, *The Age of Voltaire*, MJF Books, 1992.

12. Quoted in: Huxley, *The Perennial Philosophy*.

13. Bertrand Russell, *A History of Western Philosophy*, Simon and Schuster, 1972.

14. Edward Andrade, *Sir Isaac Newton*, Greenwood Publishing Group, 1979.

15. Quoted in: Marshall McLuhan, *The Gutenberg Galaxy*, University of Toronto Press, 1962.

16. Thomas Kuhn, *The Structure of Scientific Revolutions*, University of Chicago Press, 1996.

17. Peter Godfrey-Smith, *Theory and Reality*, University of Chicago Press, 2003.

18. This estimate was quoted in: Philip Davis and Reuben Hersh, *The Mathematical Experience*, Birkhauser, 1981.

19. Stanislaw Ulam, *Adventures of a Mathematician*, Scribners, 1976.

20. Quoted in: Nathalie Sinclair, "The Aesthetic Sensibilities of Mathematicians," in *Mathematics and the Aesthetic*, N. Sinclair, D. Pimm and W. Higginson (Eds.), Springer, 2006.

21. Godfrey H. Hardy, *A Mathematician's Apology*, Cambridge University Press, 1992.

22. See S. Chandrasekhar, *Truth and Beauty: Aesthetics and Motivations in Science*, University of Chicago Press, 1987.

23. Quoted in: Kitty Ferguson, *The Fire in the Equations*, W. B. Eerdmans Publishing Co., 1994.

24. W. S. Knickerbocker, *Classics of Modern Science*, Beacon Press, 1962.

25. Quoted in: Michael J. Mahoney, *Scientist as Subject*, Ballinger, 1976.

26. Ibid.

27. Timothy Moy, "The Galileo Affair," in *Science and Religion: Are They Compatible?*, Paul Kurtz (Ed.), Prometheus Books, 2003.

28. Mahoney, *Scientist as Subject*.

29. Ian Barbour, *When Science Meets Religion*, Harper Collins, 2000.

30. Quoted in: Robert J. Russell, William R. Stoeger and George V. Coyne (Eds.), *Physics, Philosophy and Theology*, Vatican Observatory Publications, 1988.

Part III

Facts, Theory and Mystery

Chapter 9

Describing the Indescribable

> Christian theology is always done between two poles. One pole is probably best summed up by Ludwig Wittgenstein: ... "Of that about which we can say nothing, let us be silent." ... That is an enormously important religious counsel. If God is Mystery, then let us not natter on about God like we know what we are talking about.
>
> However, this insight must be balanced by another pole – and the statement of that other pole I borrow from T. S. Eliot. Eliot wrote that ... there are some subjects about which we know in advance that anything we say will be inadequate. But these issues are so important, so crucial, that we dare not say nothing.
>
> How, then, do we talk about God, recognizing that we cannot speak of God adequately but must say something? We do what the great users of language – poets – do when trying to say the unsayable: we pile up metaphors. *Michael J. Himes* [1]

Before we can discuss the rationality of various religious teachings, we must first determine whether a scientifically minded individual can reasonably believe in a God who is both transcendent and immanent. A natural way to begin such an inquiry would be to consider what science has to say about the limits of human knowledge. This is an important question, because the existence of such limits in science would be a good indicator that there are epistemological restrictions in other fields as well. Under such circumstances, we would have to allow for the possibility that some aspects of reality will always remain beyond our reach.

If the arguments outlined above turn out to be valid, it would not be unreasonable to conceive of a God whose essence is unknowable to us. The question then becomes whether we can describe this ultimate Mystery in a meaningful way. In the opening quote of this chapter, Michael Himes warns us that any such effort is doomed to failure, but he insists that we must try nevertheless. With that in mind, in the following sections we will examine to what extent love, goodness, omnipotence (and even existence itself) constitute appropriate "divine attributes." In doing so, we will undoubtedly encounter the same linguistic problems that theologians have faced for centuries, and will have to rely heavily on metaphors and analogies. What makes our approach somewhat unique, however, is the nature of the analogies that we will use. Ours derive from the world of mathematics and physics, which ought to make them more appealing to a contemporary audience.

9.1   Unknowable Truths

As noted in Chapter 1, establishing the existence of unknowable truths is a critical prerequisite for reconciling science with religion. In the following, we will examine what mathematics and physics have to say on this subject. As a preliminary step, it will be useful to distinguish between *practically* unknowable and *intrinsically* unknowable truths. In the first case we can formulate theories and make educated guesses, but cannot hope to verify the truths experimentally. In the second case, we can only claim their existence.

Practically Unknowable Truths

To illustrate what is meant by 'practically unknowable' truths, we will consider several scientific examples that fall into this category.

Complex Systems   Chaos theory suggests the existence of relationships in complex systems that cannot be observed empirically. To see why this is so, we should first recall that in classical physics multiple experiments performed under the same conditions generally yield identical (or at least very similar) results. This is *not* the case with chaotic phenomena, where the smallest discrepancy of any kind can have dramatic consequences, and can completely alter the outcome. Given that errors are unavoidable in practice, it follows that experimental observations can provide only limited information about the behavior of complex systems.

The extraordinary sensitivity of such systems further implies that they cannot be treated in isolation. If the movements of a butterfly in California can truly affect the weather pattern in China, then accurate models of such phenomena would have to include the *entire* environment. Needless to say, developing models of this size is a formidable (and practically impossible) task. It is important to recognize, however, that this is only a part of the problem. Indeed, even if we were given a perfect mathematical description of a chaotic system, predicting its long term behavior would still require *infinite* numerical precision (see Chapter 2 for a detailed discussion). Since this is clearly not something that we can hope to achieve, it is fair to say that there are things about complex systems that will remain unknown to us. It is interesting to note that the famous physicist James Clerk Maxwell anticipated this possibility nearly a century before chaos was discovered:

"It is a metaphysical doctrine that from the same antecedents follow the same consequents. ... But it is not of much use in a world like this, in which the antecedents never again occur, and nothing ever happens twice." [2]

Quantum Mechanics There is a basic difference between the kind of uncertainty encountered in chaos and the one in quantum mechanics. In the case of chaos, we claim that the uncertainty would disappear if we had complete knowledge of the initial conditions and parameter values. This requires unlimited precision and is therefore impossible in practice, but not in theory. In quantum mechanics, on the other hand, no amount of additional information will allow us to predict what we will measure – all that Schrödinger's equation can tell us are the *probabilities* of different outcomes. Heisenberg's Uncertainty Principle complicates matters even further, since it places certain limits on the accuracy of our measurements. According to this principle, the *exact* position and momentum of a particle can never be determined simultaneously, since the states of "definite position" and "definite momentum" are mutually exclusive.

Many scientists found it difficult to accept the claim that physical reality is fundamentally unpredictable. Einstein and even Schrödinger himself openly challenged the idea that an unobserved particle is just a collection of possibilities, with no definite position or momentum. This negative attitude toward probabilistic interpretations of quantum mechanics lead to a number of "hidden variable" theories, which were based on the assumption that there is more behind quantum phenomena than we currently know. It turned out, however, that the predictions of these theories were not consistent with experimental evidence. In the early 1980s, Alain Aspect and his collaborators conclusively established that deterministic interpretations of quantum mechanics were incorrect, and that further search for hidden variables is pointless. It is now widely accepted that uncertainty is a fundamental characteristic of nature, and that our knowledge of microscopic processes is implicitly limited by this fact.

We should perhaps add that although quantum systems cannot be known in full detail, this uncertainty does *not* automatically extend into the domain of normal human experience. Such a conclusion follows from the fact that the behavior of macroscopic objects generally reflects the statistical properties of large collections of particles. The resulting mathematical models do not require a precise knowledge of the position or momentum of any given particle, and usually give rise to very reliable predictions. This is very much like the situation we encounter in thermodynamics, where it is not necessary to know the exact trajectory of each molecule of a gas in order to adequately describe its macroscopic behavior. With that in mind, it seem reasonable to assume that the limitations imposed by Heisenberg's principle have no noticeable impact on events that constitute our everyday reality.

It does not follow, however, that we should completely rule out such a possibility. Although physicists currently believe that "quantum chaos" is a rather unlikely prospect, future research may uncover mechanisms that amplify microscopic uncertainties into macroscopic phenomena (perhaps some quantum version of the "butterfly effect"). It has recently been suggested, for example, that such mechanisms might be at work in the domain of biology.[3] Proponents of this view point to the fact that DNA mutations occur on the molecular level, and are therefore subject to the laws of quantum mechanics (including, of course, the uncertainty principle). As the organism develops these random modifications are somehow "amplified," and can have very significant consequences. Given the extraordinary complexity of such phenomena, it is reasonable to expect that we may never be able to grasp all the relevant details. Nevertheless, the mere existence of such processes suggests that we should allow for the possibility that quantum uncertainties *can* affect some important aspects of the macroscopic world.

Superstrings   Although quantum mechanics and general relativity are arguably the most successful scientific theories of the 20th century, it turns out that they cannot both be completely correct. When the two theories are applied simultaneously, the resulting probabilities often tend to infinity, which obviously makes no physical sense. Superstring theory proposes a new conceptual framework in which some of these difficulties can be resolved. It is based on the assumption that the fundamental constituents of all matter (and energy) are microscopic "strings" which can be loosely viewed as vibrating one-dimensional rubber bands. Strings conceived in this way can vibrate with an unlimited number of resonant patterns, much like a guitar string of a fixed length. While the resonant patterns of the guitar string give rise to different musical notes, the vibrational patterns of microscopic strings correspond to different *particles*.

One of the main difficulties with superstring theory is a significant lack of experimental confirmation. For some of its predictions, this may merely be a matter of inadequate technology. It is conceivable, for instance, that the existence of "superpartner" particles such as *selectrons* (supersymmetric electrons) or *sneutrinos* (superpartners of neutrinos) might be established with the use of sufficiently powerful accelerators. There are other predictions, however, that are simply untestable. A typical example is the unification of gravity with the other three fundamental forces of nature (electromagnetic, strong nuclear and weak nuclear), which requires energy levels that existed only at the time of the Big Bang. It should be noted that astronomical data cannot provide us with the evidence needed to validate this claim, since the earliest history that Einstein's equations and observation allow us to study with any reliability is about one second after the Big Bang.

Another problem with superstring theories is that they require nine (or ten) spatial dimensions, which clearly contradicts all our experience. How can one interpret such a bizarre result? The explanation currently favored by most scientists points to the process of *inflation* (see Chapter 5). This hypothesis (undoubtedly a speculative one) assumes that only time and three spatial dimensions expanded, while the others remained "trapped" below 10^{-35} meters. If this were true, it would constitute another practical limit to what we can know, since we would have no access to a number of spatial dimensions that could potentially affect the microscopic world.

Cosmology   Within the field of cosmology, there are a number of theories that defy experimental testing. This is not just a matter of inadequate technology or imperfect measurement equipment, but rather a conceptual barrier which prevents empirical verification. In the following, we will briefly describe some of the obstacles that scientists have encountered in this context.

The Origins of the Universe

The Big Bang model is currently accepted as the most plausible account of how the universe was born. Its central claim is that all matter and energy originated from a *singularity*, which is a point in spacetime where the curvature and density are infinite. The equations of general relativity permit the existence of such singularities, but they do not allow us to predict what will emerge from them. This means, among other things, that Einstein's theory cannot tell us why this particular universe materialized, and not some other one. It is also unable to provide any insight into why a singularity would "expand," and ultimately turn into a universe. These are serious restrictions, which implicitly limit the explanatory power of the Big Bang theory. Why, then, do scientists take it so seriously? Most of them do so because this model makes certain *testable predictions*, which have been experimentally verified.

Given that the birth of the universe is an unrepeatable event, what kind of empirical evidence can we be talking about? On first glance, it would appear that we might be able to take a 'snapshot' of what happened, given sufficiently sophisticated instruments. Indeed, the universe has provided us with something very much like a time machine, by virtue of the fact that light travels at a finite speed (which is $c = 3 \times 10^8$ meters per second in vacuum). It is common knowledge, for example, that the light we see from the stars today originated millions of years ago. Along these lines, astrophysicist John Barrow once jokingly remarked that it is not too hard to see what the universe looked like a long time ago – the challenge is to know what it looks like now. There is a problem, however, when we attempt to study what the universe was like in the very distant past (less than one second after the Big Bang). At that time, electrons were not bound to atoms, and interacted intensely with photons (*i.e.*, light). Since this prevented the free movement of photons, no measurable data is available from that period. It is only when the universe cooled sufficiently (due to expansion) that the level of interaction decreased, allowing us to observe the effects.

The theoretical and experimental constraints outlined above suggest that the best we can do is look for *indirect* evidence of the Big Bang. The most compelling result along these lines has been the discovery of cosmic background radiation, whose spectrum was found to be in perfect accordance with theoretical predictions. This radiation, which was detected by radioastronomers Arno Penzias and Robert Wilson in 1964, is thought to be the remnant of the immense heat that pervaded the universe in the immediate aftermath of the Big Bang. Several other experimental results (such as the abundance of hydrogen and helium in the observable universe, for example) have added to the credibility to the Big Bang model, which is now a widely accepted theory.

We should note, however, that such evidence does not constitute a definitive proof that the model is correct. One can still legitimately question certain aspects of the Big Bang theory, particularly when it comes to the hypothesis that the universe originated from a singularity. Most physicists believe that singularities don't really exist in nature, and see them as a reflection of the fact that general relativity breaks down under extreme conditions. Current cosmological theories indicate that such conditions existed in the earliest moments of the universe, when temperatures presumably exceeded 10^{32} degrees (which is a staggering number by all standards).

In recent years there have been numerous attempts to develop models that can give us a better understanding of the physical processes that characterized the universe in the first 10^{-43} seconds (the so-called Planck era). Stephen Hawking has shown, for example, that when general relativity is combined with quantum mechanics, it is possible to envision a very different explanation, in which the singularity problem is circumvented.[4] Many see this result as an indication that a new theory of quantum gravity might someday resolve the mathematical difficulties that plague the Big Bang model. Perhaps they are right. Nevertheless, we should keep in mind that such a development needn't bring us any closer to determining *why* the universe came into existence. Any scientific "explanation" of this process must ultimately rely on the laws of nature which were in effect at the time of the event. But where did these laws come from in the first place? In all likelihood, this question will remain unanswered regardless of how sophisticated our theories become.

> "Even if there is only one possible unified theory, it is just a set of rules and equations. What is it that breathes fire into the equations and makes a universe for them to describe? The usual approach of science of constructing a mathematical model cannot answer the question of why there should be a universe for the model to describe."
> *Stephen Hawking*[5]

The Visible Universe

If we accept that there was a definite beginning to the universe, one of the immediate consequences is the so called 'horizon problem.' This basically means that we can see only a portion of the universe, whose limits are defined by the distance that light has traveled since the Big Bang. Until the early 1980s the horizon problem was not considered to be critical – scientists simply extrapolated that the invisible part of the universe was pretty much the same as the visible one. This has now been called into question by so-called Inflationary Theories, which allow for the possibility that the invisible universe is very different, with laws and constants of nature that have nothing to do with the ones that we observe. However, it is impossible to verify such a conjecture, since the relevant information lies in parts of the universe that are outside of our visible horizon, and will always remain so.

Black Holes

When the nuclear "fuel" of a star runs out, its outward pressure cannot counter the gravitational force any more, and its entire mass is crushed to the size of a point. This can result in a singularity of the same kind that we encountered in the Big Bang model. Within a certain region around this singularity, the pull of gravity is so strong that nothing can escape, not even light – hence the name 'black hole.' One of the immediate consequences of this property is that we can never hope to observe what goes on inside such an object.

Of course, the fact that there is no direct evidence does not preclude theoretical speculations about the physics of black holes. General relativity predicts, for example, that the matter density at the center of the black hole approaches infinity, and that time and space as we know them cease to exist (as do all known laws of physics). It is interesting to note in this context that although the model allows for a singularity to occur within a black hole, it also prohibits this singularity from influencing the outside world. Some have viewed this property as another hard limit to our knowledge (physicist Roger Penrose actually referred to it as 'Cosmic Censorship'). In the absence of such a restriction, black holes would affect the area around them, and our existing physical laws would cease to be valid. That would result in a universe which is fundamentally unpredictable to us, and science as we know it would be impossible (in fact, our very existence would be in jeopardy). It would appear, then, that in this case the unknowable becomes a necessary condition for the existence of the knowable.

Intrinsically Unknowable Truths

Both logic and science are essentially axiomatic disciplines. In science we must axiomatically posit the existence of natural laws before we can make meaningful statements about physical reality. Logic, on the other hand, has its own set of basic axioms, without which human reasoning would be impossible. With that in mind, it makes sense to speak of explanations (and therefore of knowledge) only in the context of theorems – as something that can be causally traced back to previous knowledge, and ultimately to the axioms themselves.

In justifying the axiomatic structure of scientific knowledge, empirically oriented thinkers often argue that questions such as 'Why does the universe exist?' or 'Why are physical laws the way they are?' appear to be unanswerable, and are therefore of little practical interest. From their standpoint, our understanding of nature rests on a set of fundamental "laws" to which all observations can be reduced. Since we can say nothing about the origins of these laws, we have no choice but to accept them axiomatically.

Although this "pragmatic" approach to science has a long and distinguished history, it is important to realize that it imposes certain *intrinsic* epistemological limits. Indeed, since our knowledge ends with axioms, certain questions about the nature of reality will necessarily remain a mystery to us. Restrictions of this type exist even in mathematics, where we have considerable freedom in choosing our axioms. As Gödel showed, any sufficiently complex formal system is necessarily incomplete, and will contain unprovable propositions no matter how hard we try to avoid this (see Chapter 3). This inherent "imperfection" suggests that when it comes to foundational principles, mathematics and religion really aren't all that different:

> "If we were to define a religion to be a system of thought which contains unprovable statements, so it contains an element of faith, then Gödel has taught us that not only is mathematics a religion but it is the only religion able to prove itself to be one." *John Barrow* [6]

Modern algorithmic information theory expands this outlook, and points out that the mere reduction of observations to laws is *not* equivalent to knowledge. In order to qualify as an "explanation," the laws must actually be substantially *simpler* than the data. To better understand what this means, consider a numerical sequence whose first elements are $\{1, 1.5, 1.75, 1.875, \ldots\}$. The amount of raw data in this set is obviously infinite, but its algorithmic information content is actually very small. This is due to the fact that the members of this sequence conform to a simple pattern, which can be expressed by the formula

$$s_n = \sum_{k=0}^{n} \frac{1}{2^k} \tag{9.1}$$

We can therefore legitimately assert that there is a law which "explains" the data, and allows us to predict any element in this set. A computer program that embodies such a law would require only a few lines of code, which is a quantitative reflection of its simplicity (in information theory, it is common to measure the *complexity* of a data set by the length of the shortest program needed to produce it).

Let us now contrast this example to Gregory Chaitin's constant Omega, which is a real number with infinitely many digits in its binary expansion. It turns out that computing the first N binary digits of this number requires a program whose size is very close to N bits. This effectively means that the information contained in the digits of Omega exhibits no discernible regularity, and therefore *cannot* be compressed into a simpler form. Under such circumstances the very notion of knowledge becomes questionable - the "laws" are just as complicated as the data, and it makes no sense to view them as an "explanation" of any kind.

Based on the above discussion, it is fair to say that certain mathematical truths exhibit a kind of *intrinsic unknowability*, which will not disappear with new scientific and technological advances. In the case of Omega, for example, it has been rigorously shown that we can determine only a very limited number of its bits. The remaining digits in its binary expansion elude computation, and no new theory or improvement in computing power will ever allow us to deduce them.

It is also important to recognize that the notion of intrinsic unknowability is by no means confined to mathematics and formal logic. From the standpoint of physics, for example, one could legitimately argue that our understanding of the universe is inherently limited since all our observations are necessarily views from *within*. As a result, there may well be subtle regularities and global processes that we will never be able to identify. To see this a bit more clearly, imagine that you are traveling in a windowless vehicle, at a uniform speed of 1,000 miles per hour. Based on observations made from inside the vehicle, you couldn't possibly tell whether it is moving or standing still (this is a fundamental claim of relativistic mechanics). In the absence of acceleration, only an *external* observer would be able to establish what is really going on. It is entirely conceivable that physicists face similar limitations when it comes to studying the universe as a whole.

A further constraint on our understanding of nature is related to our notion of a physical model. Over the past few centuries, science has been extraordinarily successful in studying isolated subsystems, which can be formally described by a limited number of equations and variables. Typical examples of such models can be found in Newtonian mechanics and electric circuits, where external influences tend to be insignificant and can be disregarded with only a minimal loss of accuracy. We now know, however, that certain natural phenomena *do not* lend themselves to such a description. Chaos theory, for example, shows that some systems are so sensitive to their environment that even the tiniest change can produce a significant difference in their behavior. Similarly, the non-local character of quantum mechanics suggests that any two particles which interacted at some point in the past continue to affect each other regardless of the distance between them (see the discussion of the EPR paradox in Chapter 4 for more details). In both cases, the level of connectedness is such that *everything* matters, down to the most minute detail. With that in mind, it is perfectly rational to envision the existence of phenomena whose complexity precludes any kind of modeling whatsoever. If such phenomena do, in fact, exist, it is hard to imagine how they could be analyzed or understood in any scientifically meaningful way.

> "If determinism means that the past determines the future, it can only be a property of reality as a whole, of the total cosmos. As soon as one isolates, from this global reality, a sequence of observations to be described and analyzed, one runs the risk of finding only randomness in that particular projection of the deterministic whole. Not that we have any choice. Global reality, the cosmos taken as a whole, from the most minute elementary particle to the expanding universe, is out of our reach. Science can only isolate subsystems for study, and set up experimental screens on which to project this inaccessible whole. Even if reality is deterministic, it may well happen that what we observe in this way is unpredictability and randomness." *Ivar Ekeland* [7]

What can we conclude from all this? Based on the above discussion, it seems fair to say that intrinsically unknowable truths do exist in the domain of science. A number of theoretical results clearly point to such a conclusion.

> "Gödel's incompleteness result tells us that no theory or model can be used to make all of the predictions that we would like. Chaitin's extension of Turing's Halting Problem proves that there is no reliable

way to build theories and models. Turing's incomputability shows us that simulations and natural processes may never halt, but also that we may never be able to prove this fact. Together, these three results rigorously prove that there are many things about nature that we will never be able to know with any amount of effort via experimentation, theorization and simulation. This is truly wonderful." *Gary Flake* [8]

What I find particularly interesting in Flake's statement is its final sentence. This brief thought conveys a profound sense of satisfaction, which follows from the recognition that reality will *always* contain a certain element of irreducible mystery. To me, this realization represents a natural point of contact between science and spirituality.

9.2   The Existence of God

It seems appropriate to begin our discussion of theological topics with some remarks regarding the existence of God, since this question plays a central role in all monotheistic traditions. Philosophers and theologians have debated this problem for centuries, and have often advocated very different solutions. These differences have become particularly pronounced in the 20th century, which saw the emergence of a number of philosophical movements that were openly hostile toward religion. One of the most common points of contention between secular and religious thinkers has been the claim that skepticism (and not faith) is the proper response to the absence of evidence.

> "Santa Claus is not a bad hypothesis at all for six-year-olds. As we grow up, no one comes forward to prove that such an entity does not exist. We just come to see that there is not the least reason to think that he *does* exist. ... So the proper alternative, when there is no evidence, is not mere suspension of belief .... it is *disbelief.* It most certainly is not faith." *Michael Scriven* [9]

What follows is a brief response to such arguments.

The Axiomatic Approach

Over the past millennium there has been no shortage of logical proofs for the existence of God. The arguments put forth by St. Anselm and St. Thomas Aquinas in the 11th and 13th centuries are still debated in philosophical literature, as are the more contemporary propositions based on intelligent design and religious experience [10]. A discussion of the various issues that arise in this context is clearly beyond the scope of this book. For our purposes, it suffices to say that secular thinkers have formulated some coherent and very persuasive objections to claims of this type. It is important to recognize, however, that these objections do not pose a fundamental problem for theology – they simply suggest that formal logic is not the proper basis for religious belief. This view is not foreign to the Christian tradition, whose mystical interpretations have repeatedly emphasized the fact that faith entails a *voluntary choice*. Such a choice requires a degree of uncertainty regarding God's existence, which is only possible in the absence of an indisputable logical "proof".

In light of the above arguments, a number of prominent philosophers such as Kierkegaard and Wittgenstein (and more recently Hick, Plantinga and Malcolm) have proposed a different approach, according to which formal proofs are *not* necessary for religious faith. Supporters of this view have argued that the existence of God is a basic belief, which should be accepted *axiomatically*.

> "The question is not whether it is possible to prove, starting from zero, that God exists; the question is whether the religious man ... is properly entitled as a rational person to believe what he does believe. ... Philosophers in the rationalist tradition, holding that to know means to be able to prove, have been shocked to find that in the Bible ... no attempt whatever is made to demonstrate the existence of God. Instead of professing to establish the divine reality by philosophical reasoning the Bible throughout takes this for granted." *John Hick* [11]

> "When faith begins to feel ashamed, when like a young woman for whom love ceases to suffice, who secretly feels ashamed of her lover, and must therefore have it confirmed by others that he is really quite remarkable, so likewise when faith falters and begins to lose its passion, when it begins to cease to be faith, then proof becomes necessary in order to command respect from the side of unbelief." *Søren Kierkegaard* [12]

Is such an unconditional acceptance of God's existence rational from a scientific perspective? In examining this question, it might be useful to draw an analogy with Euclid's fifth axiom. In Chapter 3 we saw that Euclid initially attempted to derive all existing knowledge about geometry from four basic axioms, which he considered to be self-evident truths. The theorems that he obtained in this way constitute what is known as *absolute geometry*. It turned out, however, that he was unable to prove the following simple proposition:

P_5 : Consider a line l and a point a that does not belong to it. Then, there is one and only one line that passes through a and does not intersect l (this would be the line that is parallel to l).

Euclid ultimately came to the conclusion that P_5 represents an obvious truth about points and lines, and decided to include this proposition as a separate axiom in his system. The set of results that followed from this assumption are commonly referred to as *Euclidean geometry*.

The status of proposition P_5 remained unresolved until the first half of the 19th century, when mathematicians Nikolai Lobachevsky and János Bolyai showed that it cannot be derived from the four original axioms. The recognition that P_5 was *unprovable* in the framework of absolute geometry formally justified Euclid's decision to incorporate it into his system. However, it also opened up the possibility that the *opposite* of P_5 could be an axiom in its own right (provided, of course, that the resulting system remains free of contradictions). This realization led to the development of several non-Euclidean geometries, whose results are anything but "self-evident."

One of the most important conclusions that we can draw from this example is that the existence of unprovable propositions presents us with two *logically equivalent* options. We can either adopt such propositions as axioms, or we can choose their opposites. In making such decisions, it is completely irrelevant whether one of the choices is less intuitive than the other. Indeed, the fact that non-Euclidean geometries seem to defy our conventional understanding of spatial relationships doesn't undermine their validity in the least. Einstein's theory of general relativity provides an excellent illustration of this point, since it uses non-Euclidean geometry to describe the structure of space and time (both of which are fundamental constituents of physical reality).

One might object to this line of reasoning on the grounds that the criteria of abstract mathematics cannot be directly applied to the real world, where beliefs ought to be consistent (at least in some measure) with human experience. This seems like a reasonable requirement, since we could otherwise axiomatically posit all sorts of strange things simply because there is no conclusive evidence against them. In addressing this objection, we should note that there is a fundamental difference between the *absence of evidence* and *logical unprovability*. In the former case, we are entitled to look for justification within the framework of standard logic, and should invoke all available scientific criteria in the process. It is therefore sensible to proportion our belief in such propositions to the *likelihood* that they are true. There are, of course, many instances where the evidence is inconclusive, and a belief may be accepted or rejected with equal justification. Superstring theory is a typical example of such a situation, since none of its claims have been verified experimentally. It is important to recognize, however, that we can reasonably expect that such issues will ultimately be settled with new advances in scientific knowledge.

In contrast to the propositions considered above, unprovable propositions *cannot* be verified in the framework of human logic. As Gödel showed, no amount of formal reasoning, speculation or experimenting will help decide such issues. It seems to me that the question of God's existence clearly falls into this category. If we agree that this is indeed the case, it is rational to accept the existence of God axiomatically, provided that the resulting beliefs do not contradict established knowledge. This last condition is critically important, since it gives a criterion for distinguishing between axioms that are permissible and those that are not. It is largely for this reason that much of the discussion in the following sections focuses on whether the Christian concept of God is consistent with the claims of science and the human experience in general.

In following this line of reasoning, we must take care not to extend the analogy too far, and treat religion as just another abstract axiomatic system. The fact of the matter is that few (if any) believers derive their faith from formal logical principles. This is, and has always been, primarily a matter of personal experience. We should also point out that the articles of faith are *not* the kind of arbitrary postulates that we normally encounter in mathematical logic. These foundational principles are, in fact, the result of the religious experience of early Christians, which has been transmitted to us in distilled form. There is no doubt that some of these claims are counterintuitive, and seem to be far removed from the world that we live in. This should not surprise us in the least, since they were inspired by extraordinary events that occurred a very long time ago, under circumstances that are quite unfamiliar to us. It is also true that they ultimately reflect the experiences of *others*, which makes them harder for us to internalize. The fact that this is so, however, does not make the articles of faith any less acceptable as the starting point of a rational belief system. As noted earlier, a logically sound system requires only internal consistency - there are no a priori restrictions regarding the choice of axioms.

We must concede, of course, that if God's existence is truly an unprovable proposition, then axiomatic atheism represents a perfectly legitimate alternative. From a logical standpoint, this is very much like saying that Euclidean geometry is just as rational as its non-Euclidean counterparts. It should be recognized, however, that in adopting a skeptical attitude toward faith one necessarily runs the risk of eliminating a whole range of possibilities. As an illustration of this point, consider an individual who refuses to trust anyone until there is incontrovertible proof of the other person's goodwill and affection. It is highly unlikely that such a standard will ever be met, given that relationships generally require a degree of axiomatic acceptance *before* they can begin to develop. The view of Christian theology is that the same holds true of our relationship toward God - a certain degree of faith and trust on our part is a necessary condition for any kind of development.

> "The rational man is the credulous man – who trusts experience until it is found to mislead him – rather than the skeptic who refuses to trust experience until it is found not to mislead him." *Richard Swinbourne* [13]

Is Existence a Necessary Attribute of God?

One can argue against the contentions of secular thinkers along different lines, by questioning whether establishing God's existence is even necessary. Philosopher Immanuel Kant claimed that 'existence' is not a valid attribute, since it tells us nothing about the entity that possesses this property. According to him, statements of existence are equivalent to number statements. If, for example, I were to say that there are 2 chairs in this room, I would be implying their existence (just like saying that there are zero chairs in this room would imply their non-existence). However, the fact that there are 2 chairs rather than 4 or 7 tells you nothing whatsoever about them. Kant used this logic to argue that existence is not a real attribute, and that God needn't formally have it to attain perfection. [14]

Another way to look at this issue is to recognize that our notion of "existence" is not limited to material objects, and can sometimes be rather imprecise. As an illustration, consider the set of all positive integers, which is commonly encountered in mathematics. In practice, we use these numbers in a very straightforward manner - it is quite obvious, for instance, what is meant when we say that we have ten fingers. It is *not* clear, however, whether the number 10 "exists" in its own right (*i.e.* without being associated with a set of physical objects). This problem becomes particularly acute if we recognize that there are some extremely large finite numbers which can be precisely defined, but cannot be related to anything in nature.[15] Can we claim that such numbers really exist? Gödel certainly believed that we can (as did some other renowned mathematicians).

> "Despite their remoteness from sense experience, we do have something like a perception also of the objects of set theory, as is seen from the fact that the axioms force themselves upon us as being true. I don't see any reason why we should have less confidence in this kind of perception, i.e., in mathematical intuition, than in sense perception. ... They, too, may represent an aspect of objective reality."
> *Kurt Gödel* [16]

> "I believe that the numbers and functions of analysis are not the arbitrary product of our spirits; I believe that they exist outside of us with the same character of necessity as the objects of objective reality; and we find or discover them and study them as do the physicists, chemists and zoologists." *Charles Hermite* [17]

When invoked in this context, the word "existence" allows for a kind of 'intuitive' knowledge which is not empirically verifiable, but is supported by a certain amount of indirect evidence. This evidence follows from the recognition that abstract mathematical concepts often lead to the discovery of unanticipated truths about physical reality.

> "In general, the case for Platonic existence is strongest for entities which give us a great deal more than we originally bargained for."
> *Roger Penrose* [18]

The notion of existence is broadened even further in the domain of quantum mechanics.

> "According to quantum rules, an unobserved system such as an atom or subatomic particle does not exist as a "real" particle. Instead it exists as a ghost-like cloud of possible physical particles. In the jargon of quantum physics, these possible physical particles are called "states" and, in general, they are the yet-to-be observable and thereby measurable attributes of our experience of any physical system. As such they reflect tendencies toward existence rather than physical existence itself." *Fred Alan Wolf* [19]

Physicists describe the dynamics of unobserved systems in terms of probability waves, which propagate through spacetime and interact much like ordinary light waves do. Since probabilities have no mass or energy associated with them, they obviously cannot be measured or detected in any other empirically acceptable way. However, this does not mean that unobserved particles are mere ideas and figments of our imagination. Unlike unicorns and goblins, they have the potential to exist *outside* the mind as well. Indeed, a quantum particle acquires *real* physical existence the moment we become aware of it. This transition is known as the "observer effect," and is one of the most perplexing phenomena in modern physics. Among other things, it seems to suggest that mind and matter cannot be completely separated, and that the subjective realm of consciousness has at least as much ontological reality as the "objective" universe. Such a view also allows for the intriguing possibility that quantum particles exist as a form of "information" which can be transformed into matter (or energy) only through the process of observation.

The precise nature of the observer effect is still a matter of considerable controversy among scientists. According to the so-called Copenhagen interpretation (a school of thought developed by Niels Bohr and his followers) the cloud of potential particles mysteriously vanishes when an observation takes place, and exactly one of the possible states is recorded. Bohr himself maintained that this process involves an inexplicable overlap of the quantum and classical perceptions of reality, which is commonly referred to as the "collapse of the wave function" (see Chapter 4 for more details).

An alternative explanation is the "many worlds" view of quantum mechanics, which was proposed by physicist Hugh Everett. This theory claims that there is no such thing as the collapse of the wave function, and that each one of the possible states actually materializes in a parallel universe. At the time of the observation the observer's mind somehow "splits" and becomes simultaneously present in *multiple universes* (which have no contact with each other). Thus, in any given world we can perceive exactly one state of the particle, but have no conscious awareness that this experience takes place on multiple fronts.

Everett's notion of "parallel existences" is quite radical, and introduces a whole range of philosophical problems. Among other things, it brings into question our understanding of freedom and moral responsibility:

> "There is no point in holding oneself responsible for one's actions and one's history if there are other 'selves' in every other possible history who have gone through every physically possible chain of events, trivial and significant, good and evil." *Mary B. Hesse* [20]

In view of the above discussion, is seems fair to say that the notion of existence is not absolute, and that its meaning depends to a large extent on the context in which it is used. Exactly how this word should be applied to God remains unclear – it may well be that in this case the usual categories of our language are simply inadequate. Theologians and mystics have recognized this a long time ago, and have often referred to God as the foundation of all other forms of existence. In the words of Paul Tillich:

> "God is being itself, not a being." [21]

9.3   The Attributes of God

Christian theology claims much more about God than mere existence, and describes Him in terms of specific attributes. This is clearly important, because it would be virtually impossible for humans to relate to a being that is completely unknowable. Without at least some information about what God is like, we would have no sense of what we should believe or what kind of conduct is appropriate. The main problem in this context is that all our concepts and words are confined to mundane experiences – we have no other frame of reference to work with. With that in mind, in this section we will examine whether it is *logically* justified to assign human attributes to God, given His transcendent nature.

The Inadequacy of Words

Virtually all religions agree that our highest experiences of God elude words. For centuries, mystics and thinkers across different traditions have stressed this fact:

> "The Tao that can be told is not the eternal Tao." *Lao Tsu* [22]

> "Almost everything that is said of God is unworthy, for the very reason that it is capable of being said." *Pope Gregory the Great* [23]

The hopeless inadequacy of language and imagery in describing God leaves us with a profound dilemma. In the opening quote of this chapter, theologian Michael J. Himes describes this dilemma as a choice between two radically different alternatives – one proposed by philosopher Ludwig Wittgenstein and the other by poet T. S. Eliot. Wittgenstein councils us to remain silent about matters of which we can say nothing, while Eliot maintains that there are certain things about which we can indeed say nothing, but dare not keep silent.

The view of T. S. Eliot seems much closer to the Christian tradition, which teaches that God is undoubtedly beyond human reason, but that we can nevertheless have some limited knowledge and experience of Him. From that standpoint, a fundamental objective of theology is to find the best ways to articulate this knowledge.

Before we discuss this issue in greater detail, it is instructive to briefly consider whether there are similar language-related constraints in science. It turns out that there are actually quite a few of them. In physics, for example, it is quite common to encounter particles such as quarks or gluons, which help us "visualize" certain relationships and allow us to make measurable predictions. It is by no means clear, however, whether such particles actually *exist*, since they have never been detected experimentally. It may well be that these entities are only useful linguistic tools for describing certain physical processes.

To this we might add the observation that the language used to describe natural phenomena constantly evolves, and occasionally allows us to envision the same physical process in multiple ways. The interaction between two electrons, for example, has traditionally been described in terms of electromagnetic forces. Such a model is intuitive, and is still commonly used in engineering practice (numerical results obtained in this manner are usually perfectly adequate). In the language of quantum field theory, on the other hand, this type of interaction is portrayed as an exchange of "virtual particles" between the electrons. These two descriptions are obviously very different, and it is fair to say that the quantum mechanical one yields more accurate predictions on the microscopic level. It remains unclear, however, whether *any* such representation properly reflects an "objective" physical reality. Many scientists have argued that this is not the case.

"Physical concepts are free creations of the human mind, and are not, however it may seem, uniquely determined by the external world." *Albert Einstein* [24]

"The word *real* does not seem to be a descriptive term. It seems to be an honorific term that we bestow on our most cherished beliefs – our most treasured ways of speaking. ... The lesson we can draw from the history of physics is that as far as we are concerned, *what is real is what we regularly talk about.* For better or for worse, there is little evidence that we have any idea of what reality looks like from some absolute point of view. We only know what the world looks like from *our* point of view. *Bruce Gregory* [25]

When seen from that perspective, scientific language does not seem to differ all that much from other forms of human communication. As in literature and theology, there is ample room for imaginative interpretations and analogies.

"It seems that the human mind has first to construct forms independently before we can find them in things. ... Knowledge cannot spring from experience alone, but only from the comparison of the inventions of the mind with observed fact." *Albert Einstein* [26]

The Via Negativa and Analogical Meaning

Christian theology has traditionally used two different linguistic models in describing God – the 'via negativa' and 'analogical meaning.' The 'way of negation' has a long history, dating back at least to the Middle Ages. Maimonides [27] and Aquinas, for example, regularly invoked negative attributes of God such as 'immutable' (not changing), 'ineffable' (not describable), 'infinite' (not limited) and 'eternal' (not subject to temporal constraints). One can legitimately ask, however, whether such descriptions tell us anything meaningful. As philosopher George Smith observes, how can we claim what God is *not* without some positive idea about what He *is*? After all, statements such as "this is not a chair" are completely vague, and tell you nothing at all about the object in question.

In examining the potential merits and weaknesses of the 'via negativa', we should first make a clear distinction between attributes that have meaningful complements and those that do *not*. It is obvious, for example, that describing an object as 'not blue' or 'not round' tells us virtually nothing about it. On the other hand, saying that it is 'not big' or 'not heavy' *does* convey some useful information (although it may not be the most accurate description). It is not difficult to show that attributes such as "infinite" belong to the latter category, although they clearly transcend our experience. Indeed, the notion of an infinite set is central to modern mathematics, and there are many explicit results that relate to its properties. We know how to compare the sizes of such sets, and have even developed a transfinite arithmetic which allows us to apply basic algebraic operations to infinite quantities.

A proper evaluation of the via negativa should also take into account the fact that certain human experiences cannot be fully grasped by the intellect. In such cases, the traditional concepts and categories that shape our thinking tend to be counterproductive, and ought to be temporarily suspended. Mystics of all traditions have held that this is a necessary condition for experiencing God:

> "[Knowing God] is not unlike the art of those who carve a life-like image from stone; removing from around it all that impedes clear vision of the latent form, revealing its hidden beauty solely by taking away." *Dionysius the Areopagite* [28]

According to Dionysius (who was an unknown 5th century mystic and author), negative claims about transcendent reality should not be viewed as literal statements of fact. Instead, they are better described as "mental guidelines" which help open us up to certain kinds of experiences. Zen koans and some forms of abstract art serve a similar purpose – they are designed to "shock" the mind out of its usual settings, and break down our conventional patterns of thought. The removal of these barriers seems to be a necessary condition for attaining certain kinds of religious knowledge.

It is interesting to note in this context that quantum mechanics has its own version of the via negativa. We often say, for example, that an unobserved particle is *not* in a state of definite position (or perhaps momentum). This is clearly a negative statement, but we have no better way of describing the underlying physical reality. The best we can do under such circumstances is to evaluate the extent to which classical concepts such as position and velocity can be applied. Unlike theology (which cannot tell us exactly how "inappropriate" our attributes are), quantum mechanics might actually have an implicit way to evaluate this discrepancy. Niels Bohr believed, for example, that this can be done by calculating a quantity know as *dispersion*, which tells us how "close" a quantum variable is to having a definite value (see Chapter 4 for more details).

The fact that science recognizes the existence of unknowable truths is a further indication that the "way of negation" is a legitimate way to describe those aspects of reality that lie out of our reach. If we assume, for example, that Inflation Theory is correct (as many cosmologists do), then we must allow for the possibility that certain portions of the universe may be governed by laws that are completely different from the ones we are familiar with. Since the physical properties of these remote regions cannot be known, negative descriptions such as "laws unlike our own" appear to be the only meaningful way to speak about them. Astrophysicist John Barrow maintains that truths of this kind may well be deeper and more important than the ones we have access to.

> "As we probe deeper into the intertwined logical structures that underwrite the nature of reality, I believe that we can expect to find more results which limit what can be known. Our knowledge of the universe has an edge. Ultimately, we may even find that the edge of our knowledge of the universe defines its character more precisely than its contents; that what cannot be known is more revealing than what can." *John Barrow* [29]

Although negation is a legitimate way of speaking about a transcendent God, Christian theology stresses the fact that He is also *immanent*. As a result, we can have some positive knowledge of the Divine, however incomplete and imperfect it may be. The key to articulating such knowledge is the concept of *analogical meaning*, which was introduced by St. Thomas Aquinas in the 13th century.

To understand the significance of this concept, we should first observe that for Aquinas words could be used in three different senses. To apply a word *univocally* to two things would be to use it in exactly the same sense. To use a word *equivocally*, on the other hand, implies two very different senses. The word 'rational', for instance, means one thing when applied to human logic, and quite another when used to describe a set of numbers.

The *analogical* meaning of words is a kind of 'middle ground' between the two extremes. According to Aquinas, this is the only way in which we can properly describe the attributes of God. Indeed, while God is wholly different from creation, there are nevertheless certain similarities, due to the fact that some part of God's essence is reflected in the world. These similarities are necessarily very limited, but they do allow us to infer something about God based on our own mundane experiences. We can therefore say that God is 'good' or 'loving' with full rational justification, although these attributes clearly mean something different when applied to a Divine being.

A number of secular thinkers have questioned Aquinas' use of analogical meaning in describing God, and have argued that such descriptions are essentially meaningless.

> "If one is to know analogically something about God, then one must know something about God literally." *William Blackstone* [30]

> "All of the supposedly positive qualities of God arise in a distinctively human context of finite existence, and when wrenched from this context to apply to a supernatural being, they cease to have meaning." *George Smith* [31]

Smith goes on to say that if you were to describe a dog as 'loyal' or 'intelligent', you would be using words analogically, in complete accordance with Aquinas' definition. However, in this case we know what is meant by the word 'dog', which is a luxury that we do not have with God. His argument, in other words, is that while analogical meaning is in itself a reasonable category, we cannot use it to speak of things that are unknowable.

In responding to such an objection, we should first recognize that God differs from finite entities in *kind*, not just in degree. This property implicitly allows us to use analogical meaning as a form of description. An illustration from mathematics may be useful in clarifying this point. In our everyday experience we regularly compare finite sets and use the attribute 'greater than' to categorize them. We perform these operations automatically, and have a clear and literal understanding of what they mean. Things become much more interesting, however, when we consider whether this attribute has an equivalent in the world of infinite sets. It turns out that it does, but its meaning is not the same. For example, the set of rational numbers and the set of integers are both infinitely large. One clearly contains the other, and yet they are said to be "equal" in size. It is difficult for us to imagine how the whole can be no larger than one of its parts, but mathematicians readily accept this fact. Along these lines, it is interesting to add that the set of real numbers is actually 'larger' than the set of rational numbers (Cantor's proof in Section 7.2 explains what is meant by this).

The above analysis shows that the attribute 'greater than' has an analogical interpretation in the context of infinite sets. There are some obvious similarities, but there is also a clear difference in kind between the two meanings. Certain properties of infinite sets simply have no counterpart in the real world. It is fair to say, then, that in dealing with infinity we can't fully grasp what we are talking about. There is no literal knowledge to rely on, since nothing in our actual experience is infinite. Nevertheless, we can draw some conclusions about infinite sets and analogically extend known concepts from the finite world. Without this kind of abstraction, there would be no calculus or modern physics.

Another illustration of analogical meaning arises in connection with the concept of a 'personal' God. Many people (particularly those of the scientific persuasion) have difficulty accepting this notion, and prefer the idea of an unknowable, transpersonal being. But what does the word "transpersonal" really mean? Does it mean that God has a personal component and much more, or perhaps that He has no such property at all? Following Aquinas' principle of analogical meaning, the former option makes much more sense. To see why this is so, imagine that you lived in a one-dimensional world where you could observe only lines and points. If that were the case, your experience of a three dimensional cube would necessarily be in the form of a line, which represents its projection. Note, however, that a cube *does* contain lines and points (as well as surfaces), so your perception would be consistent with certain aspects of reality despite its inherent limitations.

This reasoning can be extended to our perception of God, which is necessarily confined to 'projections' of His qualities. Our experience of love, goodness or personality is real, and this allows us to analogically extrapolate these concepts to God. It is therefore not irrational to claim that God does possess these attributes, although His essence is unknowable to us.

We must acknowledge, of course, that when it comes to describing God, analogical language tends to be just as limited and inadequate as any other narrative form. This is presumably what the 17th century philosopher Benedict Spinoza had in mind when he wrote that: "A triangle, if it could speak, would say that God is eminently triangular." [32] Although such conceptual restrictions apply to anything that can be said about God, we can nevertheless wonder whether certain modes of speaking are better suited for this purpose than others. Poets and theologians would probably argue that the language they use ought to be preferred, since it is rich in metaphors and analogies. Those who are scientifically minded, on the other hand, might be tempted to side with the great mathematician and physicist Hermann Weyl, who claimed that mathematics provides a natural analogical framework for conceptualizing infinity and transcendence. [33]

> "Purely mathematical inquiry in itself, according to the conviction of many great thinkers, by its special character, its certainty and stringency, lifts the human mind into closer proximity with the divine than is attainable through any other medium. Mathematics is the science of the infinite, its goal being the symbolic comprehension of the infinite with human, that is finite, means" [34]

Such a view is shared by the Russian mathematician I. R. Shafarevitch as well.

> "One is struck by the idea that such a wonderfully puzzling and mysterious activity of mankind, an activity that has continued for thousands of years, cannot be a mere chance – it must have a goal. . . Apparently, there are two possible directions. In the first place one may try to extract the goal of mathematics from its practical applications. But it is hard to believe that a superior (spiritual) activity will find its justification in the inferior (material) activity. . . I want to express a hope that . . . mathematics may serve now as a model for the solution of the main problem of our epoch: to reveal a supreme religious goal and to further the spiritual activity of mankind." [35]

In this particular case, it seems to me that both sides make a valid point. Describing God in a meaningful way is such a daunting task that we could use every symbolic tool at our disposal, and still fall short of our goal. There is, in other words, no need to choose between poetic and mathematical language - both are appropriate in this context, and we should try to combine them whenever possible. This kind of synthesis may ultimately lead to some very creative analogical interpretations of religious teachings.

9.4   Goodness, Omnipotence and Omniscience

Based on the preceding discussion, it is reasonable to assert that the concept of analogical meaning allows us to apply certain human attributes such as 'good', 'loving' and 'omnipotent' in describing the Divine nature. Although this is necessarily a limited and imperfect description, it nevertheless provides an acceptable logical framework for speaking about God in a meaningful way. The question, however, is whether the chosen attributes are consistent with human experience. For example, how can one say that God is 'good' and 'loving' with so much evil and suffering in the world? And if God cannot prevent this, how can we refer to Him as 'omnipotent'? In the following, we will take a closer look at some of the controversial issues that arise in this context.

Love and Goodness as Attributes of God

The idea that God is 'good' and 'loving' is not a universal one. In most ancient civilizations, for example, quite the opposite was true – gods were powerful, unpredictable and not particularly concerned with the well being of humanity. In connection with this subject, I am always reminded of a synopsis that I once read for Semele (one of Handel's operas based on Roman mythology). The whole libretto was vividly summarized in three words: "Gods behaving badly..."

Although contemporary views of God tend to be considerably more sophisticated than those of antiquity, many of them still remain rather unflattering. In describing God, modern secular thinkers use an array of attributes ranging from 'indifferent' and 'abstract' to 'non-existent', but the adjectives 'good' and 'loving' are conspicuously absent. Physicists Steven Weinberg and Stephen Hawking exemplify this attitude:

> "The deeper science looks into the nature of things, the less the universe seems to bear any imprint of an "interested" God." *Steven Weinberg* [36]

> "We are such insignificant creatures on a minor planet of a very average star in the outer suburbs of one of a hundred thousand million galaxies. So it is difficult to believe in a God that would care about us or even notice our existence." *Stephen Hawking* [37]

What follows is an argument that challenges such a view.

Love and the Trinity   From the standpoint of Christian theology, love is not just an attribute of God, but rather His very essence. This is the self-transcending love that is referred to as *agape*, a kind of love that goes beyond self-interest and has no ulterior motives. Simply put, it is love for its own sake, with no rewards attached to it.

The central importance of *agape* to Christianity is underscored by the fact that the Trinity itself is interpreted as a loving relationship. As early as the 5th century, St. Augustine suggested that the Father, Son and Holy Spirit may alternatively be viewed as "The One Who Loves, the Beloved, and the Love between them." [38] And since love is incomplete without others to love, it is understandable why the concept of the Trinity involves multiple persons.

Of course, it is difficult to say that anything about the Trinity is truly "understandable." Such an abstract and counterintuitive notion raises a number of conceptual questions, and is not easily reconciled with reason. One might wonder, for example, how several totally different aspects of the same being can exist simultaneously, or why Christian theology chooses to describe God as a "relationship". Making complete sense of all this is clearly an impossible task, and we must necessarily view such matters as fundamentally unknowable. Nevertheless, the concept of analogical meaning allows us to say something about the Trinity based on human knowledge and experience, without claiming to understand all the implications. Two analogies from the world of science might prove to be helpful in this context.

The first one has to do with the importance of *relationships*, which are at the heart of the theological interpretation of the Trinity. It turns out that relationships play a central role in modern physics as well. Einstein's theory tells us, for example, that there are no real "absolutes" in nature - what we see will depend to a large extent on our frame of reference. Indeed, according to special relativity, two observers who are in relative motion will generally record *different* spatial and temporal locations for the same event. The same can be said of almost any other attribute that we may use in describing an object. We may agree, for instance, that a traffic light is green at some point in time, but your green and my green will not be quite the same if I see it from the sidewalk and you happen to be driving toward it (keep in mind that colors change as we move toward or away from the object, due to the Doppler effect). It is interesting to note, however, that there is one quantity which is identical for all observers. This is the so-called "interval," which represents a mathematical measure for the spatiotemporal distance between two objects. With that in mind, it would be fair to say that the only thing we can *truly* agree on is really a kind of *relationship* between different entities. In making this point, science and theology appear to use surprisingly similar language.

> "Talk of being-as-communion has a degree of congenial consonance with modern science's way of speaking about the physical world. The advent of both relativity and quantum theory has replaced the Newtonian picture of isolated particles of matter moving along their separate trajectories in the 'container' of absolute space, by something much more interrelational in character, both in its description and behavior." *John Polkinghorne* [39]

Relationships also play a key role in the emergence of complexity. It is well known, for example, that weakly coupled subsystems behave very much like a collection of independent and unrelated entities. In such cases we can analyze the subsystems separately, and use this information to draw conclusions regarding the dynamic behavior of the overall system. However, if the interaction becomes stronger, new and qualitatively different properties might emerge which are often far more intricate than the properties of the individual subsystems. This "complexification through interaction" is closely associated with the phenomenon of self-organization, and the appearance of new forms of order in nature.

A somewhat different analogy relates to the fact that Christian theology views the Trinity as three simultaneous aspects of the same Divine being. This is not something that we encounter in our daily lives, where we tend to classify experiences in terms of exclusive opposites. To us, things are either large or small, black or white, but never both at the same time. However, quantum mechanics clearly shows that such a method is inadequate for characterizing some of nature's subtler realities. Indeed, an unobserved quantum particle is very rarely in a definite state, and is normally described as a superposition of *all* possible states. In other words, the 'state' of a quantum particle consists of many coexisting alternatives, which we can never simultaneously observe in our deterministic world. When a measurement is performed on such a particle, only a *single* state will be detected, but we cannot tell in advance which one it will be. The best that we can do in terms of prediction is to determine the probabilities of different outcomes (see Chapter 4 for more details).

The analogical connection between the Trinity and the quantum principle of superposition suggests that in their different ways, both science and theology have the capacity to reconcile and unify conflicting propositions. According Jungian psychology, this ability is essential for our spiritual well – being.

> "C. G. Jung taught that the psychic experiences that give joy and meaning in life, that indeed make life possible in any meaningful sense, involve a union of opposites, a reconciliation of opposing possibilities. *Wallace Clift* [40]

Love and Freedom   It is in the very nature of love to allow the freedom of choice. This is not only the freedom to return love, but also the freedom that allows the other person to be who they really are. Indeed, most will agree that pressuring our children to be exactly like ourselves is much closer to narcissism than to *agape*. Of course, true love doesn't imply total passivity either – it entails constraints and guidance on multiple levels. But ultimately, it is about freedom and letting be.

Christian theology claims that love is central to human existence, and that consistent development in this direction ultimately leads to *agape* and the contemplation of God. Given such a fundamental belief, the obvious question is whether we have any experiences that can confirm and reinforce it. A natural example might be the kind of unselfish love that one can develop toward another person, and subsequently toward one's children as well. Although these sentiments are often rooted in basic biological instincts (such as the survival of the species, for example), they can nevertheless be viewed as genuinely emergent phenomena which transcend their utilitarian origins. We should keep in mind, however, that the transition from instinctive behavior to selfless love is neither easy nor automatic. The realization of this possibility requires a delicate and often elusive balance between the spiritual and physical aspects of our being.

> "Evidently, eros needs to be disciplined and purified if it is to provide not just fleeting pleasure, but a certain foretaste of the pinnacle of our existence, of that beatitude for which our whole being yearns. ... There is a certain relationship between love and the Divine: love promises infinity, eternity – a reality far greater and totally other that our everyday existence. Yet we have also seen that the way to attain this goal is not simply by submitting to instinct. ... Man is truly himself when his body and soul are intimately united. ... Should he aspire to be pure spirit and to reject the flesh as pertaining to his animal nature alone, then spirit and body would both lose dignity. On the other hand, should he deny the spirit and consider matter, the body, as the only reality, he would likewise lose his greatness. ... Only when both dimensions are truly united ... is love – eros – able to mature and attain its authentic grandeur. *Pope Benedict XVI* [41]

Unselfish love needn't be restricted to personal relationships, and can be developed in other ways as well. The love of beauty, for example, is very powerful and ubiquitous among humans, and is often free from pragmatic motives. Consider the music of Bach or Mozart (or perhaps someone less classical, if you prefer) – it is something that can be deeply enjoyed but not used or possessed. The same is true of science, where one can appreciate the beauty of a theory without reference to its utility. The bottom line is that we are surrounded with opportunities to develop this kind of feeling from the day we are born till the day we die. Whether we as individuals actually do something about it is another matter. By its very definition *agape* is fundamentally different from infatuation, and is more an act of will than an emotion. As such, it must be cultivated through consistent practice, and one of the main tasks of religion is to remind us of this.

> "True love is not a feeling by which we are overwhelmed. It is a committed, thoughtful decision. The common tendency to confuse love with the feeling of love allows people all manner of self-deception. ... It is easy and not at all unpleasant to find evidence of love in one's feelings. It may be difficult and painful to search for evidence of love in one's actions. But because true love is an act of will that often transcends ephemeral feelings of love, it is correct to say, "Love is as love does." *M. Scott Peck* [42]

Goodness and Evil The preceding analysis suggests that a loving God should not micromanage the world, and must allow humans considerable freedom within certain limits (such as the laws of nature, for example). Anything else would constitute coercion, and would contradict the very notion of selfless love. We must concede, however, that where there is freedom, there is also the possibility of evil – the two appear to be logically inseparable.

Any discussion along these lines should presumably begin with the recognition that the problem of human suffering cannot be properly grasped on a purely intellectual level.

> "Suffering, our own and that of others, is an experience through which we have to live, not a theoretical problem that we can explain away. If there is an explanation, it is on a level deeper than words. Suffering cannot be "justified"; but it can be used, accepted – and, through this acceptance, transfigured." *Bishop Kallistos Ware* [43]

The view articulated above suggests that in dealing with the consequences of evil, compassion and solidarity clearly take precedence over any form of theoretical discourse. This, however, does not mean that philosophy and theology have nothing to say on the subject. There is actually a vast amount of literature that deals with the problem of evil. [44] One of the central questions in these discussions is whether moral evil is the result of human actions, God's deliberate plan, or perhaps His negligence. The so-called *free will defense* strongly endorses the first of these views, and portrays evil as the necessary price of freedom:

> "The suffering and evil of the world are not due to weakness, over-
> sight, or callousness on God's part, but, rather, they are the in-
> escapable cost of a creation allowed to be other than God, released
> from tight divine control, and permitted to be itself." *John Polk-
> inghorne* [45]

Even if we allow for the possibility that free will is the true source of all moral
evils, such arguments cannot properly account for the tragic effects of natural
disasters and deadly diseases. Events of this type are clearly beyond our control,
and their consequences cannot be attributed to human failures in any meaningful
way. One could take this line of reasoning a step further and argue that such
phenomena shouldn't be viewed as manifestations of evil, despite their negative
impact on mankind. They should instead be seen as the result of natural laws,
which are neither "good" not "bad."

The claim that nature is in some sense "neutral" is very reasonable, and is
certainly consistent with the scientific view of the world. From a theological
perspective, however, one cannot ignore the fact that the suffering associated
with natural processes appears to be at odds with the notion of a loving God
who deeply cares about human beings. A possible way to address this difficulty
would be to argue that God grants *all* of creation a certain degree of freedom,
in proportion to its nature. This would obviously mean that inanimate matter
has much less freedom than humans, but definitely enough to cause considerable
harm.

> "Exactly the same biochemical processes that enable cells to mutate,
> making evolution possible, are those that enable cells to become can-
> cerous and generate tumors ... The possibility of disease is the cost
> of life." *John Polkinghorne* [46]

John Polkinghorne refers to such a view as the *free process defense*. This
seems like a sensible approach if we agree that nature, just like human beings,
is allowed by God to be "itself." Under such circumstances, there would be no
point in assigning any kind of malicious intent to natural disasters – they are
bound to happen in a universe whose operation is autonomous.

To this we might add the observation that an external environment which
does not conform to our desires actually creates the necessary conditions for
meaningful choices and human interactions. Indeed, in order to be social beings
and moral agents, we must be able to reliably predict the consequences of our
actions. This is possible only in a lawful and "neutral" universe, whose workings
needn't always be aligned with our interests.

> "What we need for human society is exactly what we have – a neutral
> something, neither you nor I, which we can both manipulate so as to
> make signs to each other. ... But if matter is to serve as a neutral
> field it must have a fixed nature of its own. ... Again, if matter has
> a fixed nature and obeys constant laws, not all states of matter will
> be equally agreeable to the wishes of a given soul." *C. S. Lewis* [47]

As further support for Polkinghorne's "free process" argument, we should recall that matter itself is not entirely passive and possesses a certain amount of "freedom" on the microscopic level. Quantum mechanics clearly permits (and even suggests) such an interpretation.

"Speaking as a physicist, I judge matter to be an imprecise and rather old-fashioned concept. Roughly speaking, matter is the way particles behave when a large number of them are lumped together... When we examine matter in the finest detail, we see it behaving as an active agent, rather than as an inert substance." *Freeman Dyson* [48]

It is interesting to note in this context that there is an abundance of microscopic activity even in vacuum. This is a direct consequence of Heisenberg's uncertainty principle, which states that energy can fluctuate dramatically over very short periods of time. When these fluctuations are sufficiently large, they can produce particle-antiparticle pairs which are annihilated almost immediately after being born. With that in mind, it is fair to say that vacuum is actually a very dynamic place, which is "empty" only on the average.

On an even more fundamental level, Einstein's general relativity claims that spacetime itself is an active entity, which curves in the presence of matter and responds to changes in its distribution. The quantum fluctuations described above contribute to this interaction, creating tiny distortions in the fabric of spacetime which are sometimes referred to as "quantum foam." The precise manner in which this affects the observable world of our experience is still a matter of debate among scientists. We cannot, however, completely exclude the possibility that some part of the inherent quantum unpredictability may actually carry over into the macroscopic realm.

In closing this section, it is important to add that a God who is loving and personal ought to be more than just a detached observer of human suffering. Indeed, it would be logical for a being with such attributes to somehow actively participate in our physical and spiritual tribulations, and show solidarity with the plight of his creatures.

"Love makes others' sufferings its own. ... If this is true of human love, it is much more true of Divine love. Since God is love and created the world as an act of love – and since God is personal, and personhood implies sharing – God does not remain indifferent to the sorrows of the fallen world." *Bishop Kallistos Ware* [49]

This notion of a God who shares in human suffering is at the very heart of the Christian faith, and is of central importance for interpreting the doctrine of the Incarnation. The principal claim of this doctrine is that in Jesus Christ, the second person of the Trinity assumed a fully human form, while simultaneously retaining all Divine attributes. [50] According to Christian belief, through the life and death of Jesus God's involvement with human pain became very real, and His Passion remains an inexhaustible source of hope for mankind.

Explaining why an omnipotent God would choose to suffer rather than to intervene is, of course, a very difficult thing to do. This question has perplexed theologians for centuries, and is still the subject of debate. The "traditional" position of the Church in this matter (going back to the early middle ages) has been that Jesus died on the cross in order to expunge the sins of humanity. Since these sins were committed against God, they required infinite atonement, which was obviously not something that an ordinary human being could provide.

Although this belief is still quite common among Christians, it is fair to say that many contemporary theologians favor a more nuanced approach. The prevailing view suggests that Jesus suffered so that he could *share* in the pains of humanity:

> "Jesus did not pay a debt, but he redeemed an experience by sharing it." *John Polkinghorne* [51]

Such an interpretation is clearly consistent with the idea of a loving God who allows human beings to make autonomous decisions. Instead of suspending their freedom He chose to temporarily suspend His own, and experience the consequences of evil first hand.

Omnipotence

The concept of omnipotence is not easy to grasp, and raises some fundamental questions. One might argue, for example, that an omnipotent God could make anything happen, including the logically impossible. Such a claim obviously requires a response, since it seems to contradict the very foundations of human thinking. This problem was addressed quite effectively by St. Thomas Aquinas, who suggested that the term "omnipotence" excludes meaningless and illogical acts.

> "This phrase, 'God can do all things,' is rightly understood to mean that God can do all things that are possible." *St. Thomas Aquinas* [52]

The interpretation proposed above implicitly assumes that God can distinguish what is possible from what is not. Aquinas does not tell us, however, what "logical framework" God might use to make such a determination. Given that His knowledge is immesurably superior to ours, one could reasonably speculate that this would be some kind of "absolute" logic, whose mundane counterpart is only an approximation. From that perspective, human logic should be understood as a "work in progress" which is subject to change as new knowledge becomes available.

An intriguing question that arises in this context is whether we have any scientific reasons to believe that human logic needs to evolve beyond its current state. To see that this is indeed the case, it suffices to note how much modern mathematics and quantum mechanics have altered the way in which we view fundamental logical concepts. An obvious example is the so-called Law of the Excluded Middle, which states that any proposition P must either be *true* or *false*. Although this law was considered to be self-evident for centuries, we now know that there are consistent logical systems in which it is violated. In the models developed by Lukasiewicz, Post and Tarski, for instance, propositions can have three truth values, and can be classified as true, false or *undecided*. Systems of this type might perhaps be viewed as unusual and counterintuitive, but they are in no way inferior to the mode of reasoning that we employ on a daily basis. In fact, Gödel showed that such ambiguities are possible even in the domain of "conventional" logic (since there will always be statements whose truth cannot be decided in the framework of a given formal system).

> "There exists, undoubtedly, more than one formal system whose logic is feasible, and of these systems one may be more pleasing than another, but it cannot be said that one is right and the other is wrong." *Alonzo Church* [53]

A similar situation arises in quantum mechanics, where notions such as "true" and "false" take on a somewhat different meaning. To understand why this is so, we should first recall that quantum systems are inherently uncertain. As a result, if we are given a proposition such as:

P: If quantity q is measured, its value will fall in the interval $[a, b]$.

the best we can do is assign a probability to it. This, of course, doesn't prevent us from formulating a consistent logical system that conforms to the laws of quantum mechanics, but it does change the way we perceive the notion of "truth." In such a system, it would be appropriate to say, for example, that proposition P is *true* if its probability is equal to one (which we will denote in the following by $\text{Prob}(P) = 1$). If we choose to define "truth" in this manner, then we have two possible ways to define a *false* statement:

Option 1. We could say that P is false if $\text{Prob}(P) = 0$.

Option 2. We could say that P is false if $\text{Prob}(P) < 1$.

It is not difficult to show that both of these options give rise to some unusual possibilities. If we adopt the first one, there will obviously be an unlimited number of statements that are *neither* true *nor* false (these would be all propositions that satisfy $0 < \text{Prob}(P) < 1$). Such a system represents a multi-valued logic, in which statements can be *true, false* or *undecided*.

If we decide to choose Option 2, we encounter a somewhat different kind of problem. To see what this is, consider a statement that is the *logical opposite* of P (denoted $\sim$P):

$\sim$ P: If quantity q is measured, its value will *not* fall in the interval $[a, b]$.

In our conventional logical system, saying that $\sim$P is false would imply that $a \leq q \leq b$, which means that P itself must be true. It turns out, however, that this is not the case in quantum logic. As an illustration of this fact, suppose that we have established that statement $\sim$P has probability 0.2 (which would make it false according to Option 2). Since

$$\text{Prob}(P) = 1 - \text{Prob}(\sim P) = 0.8 < 1$$

must hold by definition, it follows that P *cannot* be a true statement. We have, in other words, a rather bizarre situation in which *neither* P *nor* its negation, ~P, are true statements. This is clearly very different from our intuitive notion of truth and falsity. The fact that such a possibility exists (and is consistent with certain features of the physical world) suggests that alternative logical systems should not be dismissed as "artificial." When it comes to describing certain counterintuitive features of reality, they may actually turn out to be more useful than their conventional counterparts.

The above discussion indicates that human logic is indeed a "work in progress." It also suggests that it is not unreasonable to assume that God may have chosen to restrict his actions to "what is logically possible" (as Aquinas had suggested). Having said that, however, it is important to add that we will probably never be able to properly evaluate what this means. Although we have every reason to believe that our understanding of logic will continue to evolve, it will always be limited by the fact that the human brain can process only finite amounts of information. As a result, it is likely that "absolute logic" (if such a thing exists) will remain beyond the reach of our intellect. The best we can hope for in that respect are meaningful approximations.

Omnipotence and Evil   Another common objection to Divine omnipotence has to do with the existence of evil and suffering in the world. Some thinkers have argued that an all-powerful God could have created a different world, in which humans would be free to do wrong but would never actually do so.

> "If God has made men such that in their free choices they sometimes prefer what is good and sometimes what is evil, why could he not have made men such that they always freely choose the good? ... Clearly, his failure to avail himself of this possibility is inconsistent with his being both omnipotent and wholly good." *John Mackie* [54]

In order to address this issue, it suffices to note that there may well be a number of possible worlds that God cannot realize without interfering with our free will. If He is truly committed to allowing humans to be themselves, it is entirely possible that He cannot bring about a scenario in which people always freely choose what is good. Aquinas' definition of omnipotence permits such a situation, since the laws of "absolute" logic may require God to *choose* between human freedom and the elimination of all evil. This is clearly a restriction on what God can do, but it is one that is *self-imposed*. As such, it does *not* invalidate the notion of divine omnipotence (since God makes the choice freely). [55]

To get a better sense for what this really means, imagine that you have decided to go on a diet. This choice clearly places certain constraints on what you can do, and may even have some rather unpleasant consequences in the short term. It is important to keep in mind, however, that these constraints should not be interpreted as *intrinsic* restrictions that are imposed on you by some external "law." The fact that you chose to do this freely (and could have decided otherwise) indicates that this is a case of *voluntary self-limitation*. Under such circumstances, it would not be appropriate to say that your capacity to act has been diminished in any way.

Even if we agree that God could not create a world that is wholly devoid of evil, a number of nagging questions remain. For example, if there really is a purpose to the universe (and our existence within it), why must this purpose be achieved through the cruel and painful process of evolution? In thinking about this issue, I have found it helpful to consider an analogy with a human circuit designer who creates the components and precisely specifies how they must interact. The designer is "omnipotent" in the sense that she can choose (and build) any devices that she likes, and "omniscient" in the sense that she can predict everything that will happen based on the circuit equations. However, the elements that she is working with have a nature of their own (a certain "freedom of response", if you will). As a result, once the system is activated, there is a necessary transient period until it settles down into the desired regime. In theory, it takes an infinitely long time for this to be achieved with perfect accuracy. By analogy, if we posit an omnipotent God who allows all of creation a certain measure of freedom, it is not unreasonable to assume that achieving the desired response will take time, and that the process must evolve through different stages (some of which may be quite unpleasant).

A persistent skeptic might still wonder whether God could achieve His purpose (whatever that may be) with *less* pain and suffering in the world. Indeed, it is by no means clear that the existence of *extreme* evil is necessary for the progress of humanity. One would rather expect that a good and omnipotent God could at least limit the freedom of will in individuals who are likely to commit particularly heinous acts. This doesn't seem like too much to ask - even the human circuit designer in our earlier analogy had the capability to eliminate troublesome oscillations that might arise before the desired steady state is reached.

In response to this objection, we should observe that it is clearly possible to suppress oscillations in a circuit, but *not* without changing some of its components. If we were to require this kind of adjustment on a cosmic scale, the universe would necessarily become a very different place. I think it is fair to say that we are in no position to speculate as to what this would really mean. It is very much like debating what physics would look like if its laws were altered, or if certain types of particles were to suddenly vanish from existence. One of the lessons learned from chaos theory is that in a complex system even the smallest variation can produce a fundamentally different outcome, whose features are essentially unpredictable to us. The universe appears to exhibit a similar level of sensitivity, as witnessed by the fact that even minimal changes in the fundamental constants of nature would render it inhospitable to life.

> "The laws of science, as we know them at present, contain many fundamental numbers, like the size of the electric charge of the electron and the ratio of the masses of the proton and the electron. ... The remarkable fact is that the values of these numbers seem to have been very finely adjusted to make possible the development of life. For example, if the electric charge of the electron had been only slightly

different, stars either would have been unable to burn hydrogen and
helium, or else they would not have exploded." *Stephen Hawking* [56]

Renowned physicist and theologian Ian Barbour makes a similar point regarding the rate at which the universe expanded in the immediate aftermath of
the Big Bang.

> "Stephen Hawking writes, "If the rate of expansion one second after
> the Big Bang had been smaller by even one part in a hundred thou-
> sand million it would have recollapsed before it reached its present
> size." On the other hand, if it had been greater by one part in a
> million, the universe would have expanded too rapidly for stars and
> planets to form. ... The cosmos seems to be balanced on a knife
> edge." *Ian Barbour* [57]

With that in mind, we cannot automatically assume that a more "benevolent" version of creation (or any other version, for that matter) would fulfill the
same purpose as the existing one.

It is true, of course, that we do not have a clear idea of what this purpose
might be. However, what is really at stake here is not the nature of the "solution", but rather whether or not it is *unique*. In examining this question,
we should note that our current understanding of physics characterizes the universe as a system with relatively simple laws, which generate an extraordinary
variety of outcomes. In that respect, it is almost as if nature conforms to a
minimax principle, whereby maximal complexity is achieved with a minimum
of rules. But can the same level of optimality be achieved in a different way?
Or should we perhaps agree with Leibniz that this is indeed "the best of all
possible worlds"? It seems to me that in this case we would probably do best
to suspend our judgment. In such matters we simply have no way of knowing
what is logically possible and what is not.

> "We all tend to think that had we been in charge of its creation,
> we would somehow have contrived it better, retaining the good and
> eliminating the bad. The more we understand the delicate web of
> the cosmic process, in all its subtly interlocking character, the less
> likely it seems to me that is in fact the case." *John Polkinghorne* [58]

Omniscience

The notion that God is omniscient appears to logically contradict the existence
of free will. Indeed, if we were to assume that God knows *absolutely everything* (including the future), it would be virtually impossible to claim that our
choices are truly 'free'. It would be more appropriate, instead, to say that our
thoughts and actions are *predestined*, and that human decision making is *not* an
autonomous process.

This paradox has been the subject of debate for centuries, and continues to
be an active research topic in contemporary philosophy. [59] One possible way to
approach the problem of predestination is to recognize that the prefix "pre" implicitly assumes a distinction between temporal attributes such as "before" and

"after." Although such distinctions are typical of the way humans experience time, we have no apparent reason to believe that God perceives physical reality in the same way. On the contrary, one would expect that a God who transcends spacetime should be able to see all events *simultaneously*, even though we necessarily classify them as past, present and future.

In thinking about this question, it might be useful to combine several analogical images, each of which provides us with a slightly different sense of what such a "timeless" point of view might entail. The first of these images dates back to St. Thomas Aquinas, who compared God's knowledge to what one would see when looking at a line of people from a citadel.[60] The people in the line would be ordered in a particular way, and they could clearly distinguish between those who come "before" and "after" them. From the tower, however, they would all be seen *together*, as a continuous stream of humanity. If we were now to extend this metaphor to all of creation, we could reasonably hypothesize that from God's perspective the entire history of the universe takes place in a single "eternal" present moment. Under such circumstances, time-related concepts such as prediction and predestination would appear to lose all meaning.

One could, of course, question whether it is rational to believe in the existence of a being whose knowledge transcends our usual temporal categories. This idea seems to be so contrary to our everyday experience that it is difficult to imagine. And yet, the theory of relativity clearly suggests that such speculations should not be dismissed as absurd. Indeed, it is well known that for a hypothetical observer traveling at the speed of light, the distinction between the past and future would effectively disappear, and all events would occur simultaneously (see Chapter 5 for a discussion of this property). Results such as this one ultimately led Einstein to conclude that:

> "The distinction between past, present and future is only an illusion, however persistent."[61]

The illusion that Einstein refers to appears to be unavoidable, since human beings cannot move at the speed of light (as a result of their nonzero mass). It would be fair to say, therefore, that our physical makeup precludes us from experiencing certain phenomena that our mathematical theories anticipate. It does not follow, however, that such constraints must apply to God, who is presumably *immaterial*. With that in mind, we can conclude that thinking of God as a being whose knowledge extends beyond time may perhaps be counterintuitive, but is certainly not irrational.

It is interesting to note in this context that Einstein's understanding of time is remarkably similar to the theologically inspired theories proposed by St. Augustine in the 5th century. In Book XI of his *Confessions* Augustine wrote:

> "What, then, is time? If no one asks me, I know; if I wish to explain to him who asks, I know not."[62]

Augustine attributed "reality" only to the present moment, although he acknowledged that we subjectively distinguish between three kinds of temporal experience:

"There is a present of things past, a present of things present, and a
present of things future. ... The present of things past is memory;
the present of things present is sight; and the present of things future
is expectation." [63]

In interpreting this statement, we should bear in mind that Augustine treated
both memory and expectation as psychological mechanisms that are located in
the *present*. According to him, our experience of the past, present and future is
not an accurate reflection of reality, and is instead a consequence of our mental
limitations.

Even if we agree with Augustine, and concede that the problem of predestination is ultimately a result of our inability to properly grasp the notion of
time, this still tells us nothing about how a God with "timeless" knowledge
might interact with his creation. It remains unclear, for example, whether his
omniscience is limited in any way, and whether it interferes with our ability to
make free and uncoerced choices. It is also difficult to envision how a God who
transcends the spacetime continuum could experience change, given that change
is essentially a time-related concept. Both of these questions are of crucial importance to Christian theology, since developing any kind of "relationship" with
God requires both free will and a certain level of reciprocity.

In attempting to resolve these difficulties, it might be helpful to move beyond
the traditional image of a "static" deity that has pervaded theological literature
for the past two millennia. What we have learned about complex systems and
strange attractors in recent years suggests a rather different metaphor:

"Just as physics can conceive of an equilibrium which is not simply a
static staying put but which is the dynamical exploration of a pattern
of possibility, so we can surely conceive a dynamical understanding
of perfection, which resides, not in the absence of change, but in
perfect appropriateness in relation to each successive moment. It is
the perfection of music rather than the perfection of a statue." *John
Polkinghorne* [64]

If we adopt Polkinghorne's perspective, it would be entirely reasonable to
assume that creation is, in fact, a "work in progress." Such a view finds considerable support among contemporary theologians, many of whom are sympathetic
to the idea that creation is a *continuous* and still unfinished process. It would be
interesting to see, however, whether this outlook is consistent with the explanations currently favored by modern physics. In order to make such an evaluation,
we should first recall that the standard theoretical model used to describe the
universe involves a single four dimensional "block" of spacetime. Within this
"block," each object is represented by a unique trajectory (or *world line*), which
starts at the point in spacetime where the entity in question comes into being,
and ends where it ceases to exist. It is important to keep in mind in this context
that when spacetime is viewed as a whole, the world lines of different objects
appear in their *entirety*. In other words, our current mathematical models of the
universe do not assume that the history of a physical object gradually emerges
over time. The "flow" of time actually doesn't enter into this picture at all, and

physicists prefer instead to represent the past, present and future of any given object in a single "snapshot," as a complete trajectory.

The theoretical model outlined above reinforces the view that our intuitive perception of time is probably *not* an adequate representation of reality. It also suggests that an observer who could view spacetime as a "block" would see the entire history of the universe at a single glance. If we were now to analogically extend these ideas to theology, we could easily envision God as a "timeless" observer, who possesses knowledge of the past, present and future of every particle, object and living being that ever existed. Such a "mental image" of God has some obvious advantages, since it allows us think of Him as an omniscient being while preserving the established scientific framework for describing the universe.

We should be careful, however, not to take this line of reasoning too far. The possibility that God knows everything about our future actions brings into question fundamental theological notions such as free will and human autonomy, both of which are essential to the Christian understanding of faith and moral responsibility. In order to avoid the logical difficulties that are associated with such an "unconstrained" interpretation of divine omniscience, it will be necessary to make certain modifications to our previous analogy. A natural way to do this would be to recognize that the "block" model of the universe may just be a convenient mathematical representation, which needn't be an accurate reflection of physical reality on the cosmic scale. Based on such considerations, the prominent biochemist and theologian Arthur Peacocke has argued that it is impossible to tell whether world lines *really* extend beyond what we perceive as the "present moment." [65] If we were to adopt this point of view, it would not be unreasonable to propose a somewhat different "mental image" of God's interaction with the universe, in which some world lines are "unfinished" and the "future" portion of spacetime has yet to be created.

What would this mean from a theological perspective? To begin with, it would imply that God possesses timeless knowledge of all *created* events and entities (*i.e.*, all that *is*). The uncreated future, however, would remain open and unknowable even to God. Given His superior intelligence, God could perhaps estimate the probabilities of different outcomes, but the actual course of events would *not* be predetermined (due in part to the existence of human free will). Such an interpretation also suggests that what transpires in the physical world could conceivably influence how God creates the next "slice" of spacetime. The interactive nature of this process would allow God to engage His creatures *within* time, and give them the opportunity to play a role in shaping the universe. This would, of course, entail certain restrictions on God's omniscience, in the sense that His knowledge would be limited to what has been *created so far*. One could reasonably argue, however, that this restriction is "self-imposed," since God freely chose such a course of action.

How might we describe this creative interplay between God and the physical universe (including human beings, of course)? The relationship would undoubtedly be a highly *asymmetric* one, since it involves the interaction between finite and infinite entities. [66] Nevertheless, it is not impossible to envision how such a process might unfold. To see how that can be done, we should first recall that scientists and engineers deal with mathematical infinity on a regular basis, and have developed some sophisticated techniques for this purpose. Although these techniques were clearly not designed with theological questions in mind, they can still provide us with some meaningful insights and analogies.

A typical example of such an analogy arises in the context of electric power systems, where we often use the concept of an "infinite bus" to represent a generator with an unlimited capacity to deliver or absorb energy. It is not difficult to show that such an idealized generator influences the system dynamics without being affected itself, since its infinite inertia precludes any form of measurable temporal change. This means (among other things) that all the variables used to describe such a generator necessarily remain *constant* throughout the interaction.

If we were to extend this line of reasoning to theology, the logical conclusion would be that we cannot expect to observe any variations in the "variables" (*i.e.,* attributes) that we use to characterize God. From our perspective, His "state" (interpreted perhaps as His disposition toward humanity) would remain fixed regardless of what we do. Having said that, however, we should add that there is no a priori reason to assume that God himself is incapable of experiencing change. The fact that we cannot register this may simply indicate that the concept of "change" has a very different meaning when applied to an infinite being.

What kind of "change" might we be talking about in this case? While such a question clearly eludes a precise answer, we can perhaps outline some of the possibilities by drawing an analogy with set theory. This area of mathematics is particularly suitable for our purposes, since it recognizes that there are many different types of infinity. We know, for example, that the set of integers and the set of real numbers both contain infinitely many elements, but the latter is considered to be "larger" in a certain well-defined sense. We can even construct an entire hierarchy of infinite numbers, each of which is "larger" than all the preceding ones. If we were now to analogically extrapolate these notions, one could reasonably argue that an infinite being might have multiple "states," and that it could possibly transition from one to another under some circumstances. It goes without saying, of course, that these transitions would necessarily remain imperceptible to us, since our senses (and our instruments as well) cannot detect such variations. This implies that we are in no position to empirically determine whether our thoughts and actions have any effect on God. It is entirely possible, however, that we may be able to experience this interaction in a different and far more intuitive way, which is neither sensory nor quantifiable. This is, in fact, precisely how mystics of all traditions have described their encounters with God.

9.5   Notes

1. Michael J. Himes, "Finding God in All Things: A Sacramental Worldview and Its Effects," in *As Leaven in the World: Catholic Perspectives on*

Faith, Vocation and the Intellectual Life, Thomas Landy (Ed.), Rowman and Littlefield, 2001.

2. Quoted in: Kitty Ferguson, *The Fire in the Equations*, W. B. Eerdmans Publishing Co., 1994.

3. Thomas Tracy, "Evolution, Divine Action and the Problem of Evil," in *Evolutionary and Molecular Biology*, R. J. Russell, W. R. Stoeger and F. J. Ayala (Eds.), Vatican Observatory Publications, 1998.

4. Hawking's proposal is highly speculative, and is most probably impossible to verify experimentally. Among other things, this model postulates the existence of "imaginary time," which is fundamentally different from the time that we experience and are able to measure.

5. Stephen Hawking, *A Brief History of Time: From the Big Bang to Black Holes*, Bantam Press, 1988.

6. John Barrow, *Pi in the Sky*, Little, Brown & Company, 1993.

7. Ivar Ekeland, *Mathematics and the Unexpected*, University of Chicago Press, 1988.

8. Gary Flake, *The Computational Beauty of Nature*, MIT Press, 2000.

9. Michael Scriven, *Primary Philosophy*, McGraw-Hill, 1966.

10. There is a wealth of literature on this subject. A good starting point might be Pojman's anthology: Louis P. Pojman, *Philosophy of Religion*, Wadsworth, 2003.

11. John Hick, *Arguments for the Existence of God*, Macmillan, London and Basingtoke, 1971.

12. Søren Kierkegaard, "Subjectivity is Truth" in Pojman, *Philosophy of Religion*.

13. Richard Swinbourne, *The Existence of God*, Oxford University Press, 1979.

14. This was part of Kant's critique of St. Anselm's ontological proof for the existence of God. For further details, see *e.g.* Brian Davies, *An Introduction to the Philosophy of Religion*, Oxford University Press, 1993.

15. The number *mega*, which was introduced by Polish mathematician Hugo Steinhaus, is so large that it can only be described in terms of a construction process.

16. Quoted in: Philip Davis and Reuben Hersh, *The Mathematical Experience*, Birkhauser, 1981.

17. Quoted in: Bruce Gregory, *Inventing Reality*, John Wiley and Sons, 1990.

18. Quoted in: Barrow, *Pi in the Sky*.

19. Fred Alan Wolf, *Mind Into Matter*, Moment Point Press, 2001.

20. Mary B. Hesse, "Physics, Philosophy and Myth," in *Physics, Philosophy and Theology*, R. J. Russell, W. R. Stoeger and G. V. Coyne (Eds.), Vatican Observatory Publications, 1988.

21. Paul Tillich, *Systematic Theology*, University of Chicago Press, 1971.

22. Lao Tsu, *Tao Te Ching*, Vintage Books, 1989.

23. Quoted in: Will Durant, *The Age of Faith*, Simon and Schuster, 1971.

24. Quoted in: Gregory, *Inventing Reality*.

25. Ibid.

26. Ibid.

27. Maimonides (1135-1204) was a major medieval Jewish scholar.

28. Dionysius the Areopagite, *On the Divine Names* and *Mystical Theology*, Nicolas-Hays, Inc., 2004.

29. John Barrow, *Impossibility: The Limits of Science and the Science of Limits*, Oxford University Press, 1999.

30. William Blackstone, *The Problem of Religious Knowledge*, Prentice Hall, 1963.

31. George Smith, *Atheism: The Case Against God*, Prometheus Books, New York, 1989.

32. Quoted in: Will Durant, *The Story of Philosophy*, Simon and Schuster, 1972.

33. It is important to keep in mind that there is a fundamental distinction between mathematical and metaphysical infinity. The latter concept belongs in the domain of philosophy and theology, and cannot be properly examined in the framework of mathematics.

34. Quoted in: Davis and Hersh, *The Mathematical Experience*.

35. Ibid.

36. Steven Weinberg, *Dreams of a Final Theory*, Pantheon Books, 1992.

37. Quoted in: Ferguson, *The Fire in the Equations*.

38. Gareth Matthews (Ed.), *Augustine: On the Trinity*, Cambridge University Press, 2002.

39. John Polkinghorne, *The Faith of a Physicist*, Fortress Press, 1996.

40. Wallace Clift, *Jung and Christianity*, Crossroads, 1986.

41. Pope Benedict XVI, *Encyclical Letter "Deus Caritas Est,"* 2005.

42. M. Scott Peck, *The Road Less Traveled*, Simon and Schuster, 1998.

43. Bishop Kallistos Ware, *The Orthodox Way*, St. Vladimir's Seminary Press, 1995.

44. A great deal of literature has been devoted to this issue. A good starting point might be: Marilyn Adams and Robert Adams, *The Problem of Evil*, Oxford University Press, 1990, as well as: Daniel Howard – Snyder (Ed.), *The Evidential Argument from Evil*, Indiana University Press, 1996.

45. John Polkinghorne, *Quarks, Chaos and Christianity*, Crossroad, 2000.

46. Ibid.

47. C. S. Lewis, *The Problem of Pain*, Harper Collins, 2001.

48. Freeman Dyson, *Infinite In All Directions*, Harper and Row, 1988.

49. Bishop Kallistos Ware, *The Orthodox Way*.

50. The Council of Chalcedon held in 451 A.D. describes Christ as *one person* with *two natures* – human and divine.

51. Polkinghorne, *The Faith of a Physicist*.

52. St. Thomas Aquinas, *Summa Theologiae*, Part I, Q. 25.

53. Quoted in: Barrow, *Pi in the Sky*.

54. John Mackie, "Evil and Omnipotence" in *Philosophy of Religion*, Louis Pojman (Ed.), Wadsworth, 2003.

55. A thorough discussion of this issue can be found in: Alvin Plantinga, *God, Freedom and Evil*, Harper and Row, 1974.

56. Hawking, *A Brief History of Time*.

57. Ian Barbour, *When Science Meets Religion*, Harper Collins, 2000.

58. Polkinghorne, *The Faith of a Physicist*.

59. A thorough discussion of this topic can be found in: John Fischer (Ed.), *God, Foreknowledge and Freedom*, Stanford University Press, 1989.

60. St. Thomas Aquinas, *Commentary on Aristotle's Peri Hermeneias*, I, Lectio 14.

61. Quoted in: Paul Davies, *God and the New Physics*, Simon and Schuster, 1983.

62. St. Augustine, *Confessions*, Knopf Publishing Group, 1998.

63. Ibid.

64. Polkinghorne, *The Faith of a Physicist.*

65. Arthur Peacocke, *Paths from Science Toward God*, Oneworld, Oxford, 2001.

66. When describing God as an "infinite entity," we are, of course, using this term in its metaphysical sense.

Chapter 10

Four Difficult Questions

"Science leads us to hope that complete understanding is potentially within the grasp of human collective reason, but science is not overly confident of finding it. ... Religion is far more optimistic than science that in some manner beyond our present concept of human reason, we can know 'everything important.' Perhaps the most significant difference between science and religion is that science thinks that on this quest we are entirely on our own. Religion tells us that although we who seek the truth may ride imaginary horses, Truth also seeks us." *Kitty Ferguson* [1]

In this chapter, we will focus on four questions that have been particularly controversial in the debate between science and religion. Not surprisingly, we will begin with the problem of miracles. Given the fundamental role that such events play in the Christian tradition, it is important to examine whether they explicitly contradict what we know about the physical world. If that turns out to be the case, some of the central beliefs in Christian theology would be brought into question.

After tackling this difficult issue, we will turn our attention to theology itself, and its ability to uncover truths about the nature of reality. Over the past few centuries, secular thinkers have repeatedly asserted that religious teachings lack objectivity, and that theology has no clear criteria for distinguishing truth from falsehood. In order to respond to such claims, we must examine the theological method of inquiry, and consider how it compares with the modern scientific paradigm.

The third question that we will address in this chapter has to do with the uneasy relationship between religion and evolution theory. There has been a great deal of controversy on this subject ever since Darwin published his classic book *On the Origin of Species* in 1859. Our objective in the following will be to establish whether evolution theory can somehow be reconciled with the notion of a "divine purpose," and whether we have any scientific grounds to assign a special role to humans in this process.

In the final section, we will address the politically charged problem of religious diversity, and inquire whether it makes sense to claim that there is a single path to the "truth." At some point in their development, almost all major religions have promoted such an exclusive outlook. Not surprisingly, this inherent intolerance for the views of other traditions has lead to a great deal of human misery. This is undoubtedly one of the main reasons why some secular thinkers tend to view religion as a potentially harmful social phenomenon. Our task in the following will be to consider whether interfaith dialogue is a viable option, and evaluate whether science can perhaps play a mediating role in the process.

10.1   Are Miracles Possible?

Of all the major religious traditions, Christianity probably has the most at stake when it comes to miracles. The Resurrection, for example, represents a central part of its belief system, and the entire teaching of the Church rests on this foundation. Christian theology also views miracles as an important source of religious knowledge:

> "God has probably blessed us as much through the air we breathe as through other gifts; but if piety had to wait for people to infer God's goodness from the availability of oxygen, it would have been long in coming." *Huston Smith* [2]

Secular thinkers have a very different perspective. To them, miracles defy statistics, natural laws and pretty much all of human knowledge. As such, they are no more than myths which can be explained as products of ignorance and superstition.

Defending the possibility of miracles in the face of such opposition (and indeed our own experience as well) is a formidable task. One can argue, of course, that these are profound mysteries which defy our understanding, and that we would do best to remain silent on the subject. But the real challenge is to say more, and perhaps provide some rational justification for belief in miracles.

Before entering into such a discussion, it might be useful to provide a working definition for the word "miracle." In the following, we will use this term to describe events of religious significance whose cause can be reasonably attributed to God. These events must necessarily possess a "supernatural" dimension, in the sense that they represent aspects of reality that lie beyond the reach of science. It is important to recognize in this context that the reality we are referring to is ultimately the *same* one that science investigates. The distinction that we draw between "natural" and "supernatural" phenomena is primarily a cognitive one, and the boundaries separating these two attributes are defined by the intrinsic limits of human knowledge.

In thinking about miracles, we should also keep in mind that the manner in which we perceive unusual events is significantly influenced by our basic beliefs about reality. Indeed, when interpreting unprecedented experiences our responses are necessarily limited to explanations that we consider to be physically and logically possible.

"The question whether miracles occur can never be answered simply by experience. ... If anything extraordinary seems to have happened, we can always say that we have been the victims of an illusion. If we hold a philosophy which excludes the supernatural, this is what we always shall say. What we learn from experience depends on the kind of philosophy we bring to experience. It is therefore useless to appeal to experience before we have settled, as well as we can, the philosophical question." *C. S. Lewis*[3]

In view of these observations, in the following we will examine whether belief in miracles is a rational possibility, and whether the scientific and religious points of view can somehow be reconciled in this matter.

Miracles as a Result of Ignorance

Generations of Eastern Europeans (myself included) grew up with Karl Marx's famous line that "Religion is the opiate of the masses." When this statement was explained in textbooks, miracles were usually among the first examples to be invoked. Their appeal to broad segments of the population was attributed to a lack of proper education, and a deliberate suppression of scientific facts by the Church.

This is a very potent argument, regardless of one's religious views.

"It is doubtful that anyone, even the most devout Christian, actually believes that as mankind advanced in knowledge, the frequency of miracles coincidentally tapered off." *George Smith*[4]

Thinkers of this persuasion tend to view 'miracles' as events that cannot be explained by presently known laws, but they fully expect that all such occurrences will be accounted for at some point in the future. After all, history abounds with examples where initially mysterious phenomena were subsequently explained by new scientific findings.

It would, of course, be absurd to deny that there is a lot of truth in the criticisms articulated above. There has indeed been an abundance of ignorance, superstition and irrationality throughout human history, a fact that prompted Einstein to say that:

"Only two things are infinite, the universe and human stupidity, and I'm not sure about the former." [5]

Even today, significant numbers of practicing Christians routinely engage in activities that can best be described as pagan. It is also quite true that modern science has dispelled many myths and false beliefs about the way nature works. But this does not mean that all speculation about miracles must automatically be dismissed. Indeed, one of the most remarkable insights of modern science is the realization that there are many things that will remain unknowable to us by their very nature.

"A paradoxical revelation that we can know what we cannot know
is one of the most striking consequences of human consciousness.
Advanced formulae and theories predict that there are things they
cannot predict." *John Barrow*[6]

In view of that, the expectation that all "miraculous" events will some day
be explained appears rather unrealistic. It seems to me that a certain element of
mystery is an unavoidable part of the human experience, and that this element
cannot be removed by a simple accumulation of knowledge.

The contention that belief in miracles is an outgrowth of human ignorance
is sometimes also justified by claims that purported "supernatural" events are
actually no more than delusions or hallucinations.

"From a scientific point of view, we can make no distinction between
a man who eats little and sees heaven and the man who drinks much
and sees snakes." *Bertrand Russell*[7]

Although one can certainly appreciate Russell's sense of humor, the fact remains that such comparisons between alcoholics and mystics aren't particularly
persuasive. Indeed, wouldn't the experience of an infinite being necessitate an
unusual state of mind? In pondering this question, we should keep in mind
that the words 'unusual' and 'incorrect' are by no means synonymous. After all,
many great scientists and artists have exhibited eccentric behavior, but their
claims were nevertheless perfectly valid.

It is true, of course, that certain emotions (or images) can be externally
stimulated by drugs or alcohol. However, it is equally true that there is nothing
permanent about them, and that what follows is necessarily detrimental to both
physical and mental health. In contrast, mystical experiences have often led
to profound and lasting transformations of character. When people respond to
such experiences by dedicating their lives to the well-being of others, there is no
doubt that something fundamentally positive has occurred. If we agree, then,
that extraordinary phenomena should be judged by their fruits, it is obvious
that mystical visions and "delirium tremens" belong to very different categories.

The Unlikelihood of Miracles

Ever since the dawn of the scientific age, miracles have been repeatedly dismissed
on probabilistic grounds. By the end of the 18th century, it was quite common
to encounter statements such as:

"No testimony for any kind of miracle has ever amounted to a probability, much less to a proof." *David Hume*[8]

or

"Is it more probable that nature should go out of her course, or that
a man should tell a lie?" *Thomas Paine*[9]

In response to arguments of this type, we should first consider what is commonly understood as a 'law' of nature. To that effect, let us suppose that we have a hypothesis P whose validity would imply observations $\{q_1, q_2, \ldots, q_m\}$ as its necessary consequences. If all of these observations were actually verified experimentally, most scientists would be inclined to accept P as a valid 'law'. It is important to recognize, however, that although this kind of reasoning can be justified *statistically*, its logical implications are rather limited. This is due to the fact that there could be any number of alternative hypotheses which entail the same set of observations as their necessary consequences.

The above example suggests that the only proper way to test the logical validity of a 'law' would be to attempt to *disprove* it. Indeed, if we were to conclusively determine that at least one of $\{q_1, q_2, \ldots, q_m\}$ does *not* arise in nature, then we could be certain that the law is incorrect. From the standpoint of propositional logic, this is the *only* kind of indisputable conclusion that experimental evidence can produce. It is interesting to note, however, that despite its sound logical foundation, such a methodology is not very appealing to the human mind. A number of psychological studies performed over the past few decades have shown that most individuals (including scientists) seem to clearly prefer confirmatory reasoning, where the focus is on discovering evidence that supports the hypothesis in question. [10] This insight is actually not a particularly new one – Francis Bacon arrived at the same conclusion some 400 years ago:

"It is the peculiar and perpetual error of the human intellect to be more moved and excited by affirmatives than by negatives." [11]

The bias in favor of affirmative evidence is perhaps most apparent in the case of the Riemann hypothesis (see Chapter 3), where the Clay Mathematics Institute currently offers a one million dollar prize for a proof that this conjecture is correct. Interestingly, no reward is offered for a disproof, although this would be an equally relevant mathematical result.

In view of the above discussion, it seems reasonable to say that a hypothesis can be referred to as a 'law' if no exceptions to its necessary consequences have been observed (such a definition is certainly consistent with accepted scientific practices). But if we do that, we must then concede that laws conceived in this manner *cannot* claim logical certainty. Bertrand Russell, himself a great logician, wittily made this point by observing that: "It is not certain that I will die, only extremely likely." [12] He lived to be 98 years old, and nearly proved this statement empirically!

If we agree, then, that laws of nature are essentially statistical, the worst we could say about miracles is that their probability is negligibly small. But this is *not* equivalent to saying that such events are impossible. To gain a better understanding of what this really means, let us consider a simple thought experiment in which I think of an integer and ask you to guess what it is. The probability of your success will obviously be arbitrarily close to zero, since there is only one favorable outcome and an unlimited number of choices. And yet, it is clearly *possible* that you may guess correctly.

As a further illustration of this property, consider the fact that our individual existences are extremely unlikely events, given the odds of all our ancestors meeting and procreating. Actually, whenever we factor in all the plausible scenarios that *didn't* happen, any particular event has a negligible chance of occurring. This line of reasoning suggests that statistical interpretations are often inadequate when it comes to complex phenomena. Indeed, based on probabilities alone, the fact that you are alive and reading these lines constitutes a genuine "miracle."

Of course, the realization that extraordinary events cannot be ruled out logically is not likely to sway too many skeptics. Since miracles are supposed to take place in the world of our sensory experience, it would appear that the real challenge for theists is not to establish the logical possibility of miracles, but rather to show that we can conceive of physical laws that can occasionally be 'broken' on the macroscopic level. To see whether this can be done, let us consider a chaotic system which is on the borderline of intermittency (that is, very close to the bifurcation point). The model for such a system is as deterministic as Newton's laws – there is no trace of uncertainty in the mathematical description. Under such circumstances, one would expect to encounter reasonably predictable dynamics. However, Fig. 10.1 shows that such a system is characterized by long periods of regular behavior, which are interrupted by sudden aperiodic bursts.

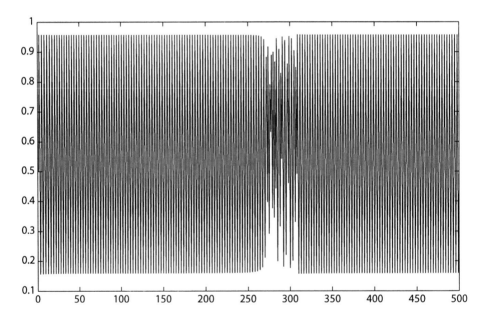

Fig. 10.1. Intermittent behavior of a chaotic system.

If we assume that this phenomenon represents a class of natural "laws," what would that really mean? To begin with, it would imply that there are macroscopic laws which apply 'almost always.'[13] The underlying equation obviously allows for such a possibility, although it cannot predict when the aperiodic activity will occur. Chaos theory also claims that such events are *unique*, and that the system will behave in the "usual" way both before and after. From an empirical standpoint, it is conceivable that entire generations of observers could register only the regular dynamic pattern, with no idea that other forms of behavior are possible. Even if an occasional anomaly were to be recorded, it would probably be dismissed as a measurement error (or it could perhaps be attributed to the mental instability of the observer).

If we do allow for the existence of physical laws that can be 'broken' sporadically, wouldn't this imply that 'miracles' are just a special class of natural phenomena? Perhaps, but this certainly doesn't rule out the possibility of Divine involvement in such events. It is reasonable to assume, for example, that a God who loves his creation would choose to work *through* the laws of nature, rather than against them. Indeed, the notion of a God who violates the rules he himself created is rather unappealing from a theological standpoint.

> "This approach suggests the image of the cosmic magician, who creates a flawed universe and prods it whimsically from time to time to correct the errors." *Paul Davies* [14]

It would appear, then, that the possibility of occasional empirical irregularities in nature fits in very nicely with the Christian idea of God. In evaluating the merits of such a claim, we should also keep in mind that science cannot explain *why* laws like the one described in Fig. 10.1 exist, or how prevalent they are in the universe. The most we can say is that complex systems allow for the possibility of unprecedented events, and that Divine involvement cannot be automatically ruled out as a contributing factor. Under such circumstances, it is quite possible that we may have to change our conventional understanding of physical laws, and reevaluate the extent to which they are binding.

> "Do chaos and complexity theory allow us to reconcile the idea of an intervening God with a rational universe? ... The old dichotomy between Intervening God and a Clockwork Universe has crumbled. The picture that is emerging is subtle and complicated in ways we are only dimly beginning to understand. In a universe which combines predictability and freedom, as these theories suggest ours does, insisting there was a violation of the fundamental laws of the universe if the waters parted for the Israelites or even if Christ rose from the dead might be tantamount to insisting that the Constitution of the United States is violated if traffic is allowed to flow the wrong way on a one-way Fifth Avenue for several hours to accommodate a St. Patrick's Day parade." *Kitty Ferguson* [15]

Lack of Empirical Evidence

The discussion in the previous section suggests that miracles are logically possible, and are perhaps even consistent with certain types of physical laws. But what about the *evidence* for their existence? Is it reasonable to require that claims of miracles be empirically testable before they are accepted? According to philosopher Paul Kurtz:

> "[Miracles] have an empirical component and are not completely transcendental, and hence they are capable of some experimental testing and historical reconstruction of their claims." [16]

It seems to me that this is a sensible requirement, and that some form of empirical evidence is indeed necessary. However, the standards of proof cannot be the usual ones applied in science (repeatable experiments, observation etc.), since these events are *unique*. As such, they can perhaps be compared to *singular* phenomena such as the Big Bang. It is well known that modern physics has no formal mechanism for dealing with these events, and that we cannot conclusively establish whether or not they really occurred - the best we can do is look for detectable "traces" that they may have left. This is precisely what scientists have done in evaluating the Big Bang hypothesis. The key evidence in this case is the existence of "background radiation," which is believed to be a remnant of the original explosion.

If we were to extend this criterion to the domain of religion, it could be argued that miracles ought to be judged by the "traces" they left in human history (measured, perhaps, by the number of lives that were permanently transformed by these events).

> "Christianity rests its case on the evidence of the astounding trans-forming power we see manifested in human lives, power that makes it possible for individuals not only to change the world around them, but, even more surprisingly, to change themselves in ways we have no right to expect humans to be able to change themselves." *John Spong* [17]

The difficulty with this approach is that the evidence usually comes in the form of historical testimony, which isn't always a reliable source of information. Most historians believe, for example, that ancient texts tend to be ambiguous, and often fail to separate facts from symbols and metaphors. One could therefore reasonably conclude that such data rarely meets the objective standards set by the natural sciences.

It is important to recognize, however, that these standards are *not* absolute. Following the work of Norwood Hanson [18] in the late 1960s, both philosophers and psychologists have questioned the impartial objectivity of science, and have pointed out that it is often the theories themselves that determine what data we will collect. The argument here is that scientists (like all other fallible humans) tend to look for what they expect to see, and often dismiss (or simply fail to detect) data that doesn't fit in with the current paradigm. When such data is recorded, it is usually labeled "erroneous" or "coincidental", and becomes acceptable only when the paradigm changes. [19]

Questions regarding the limitations of objective data gathering have been raised within the scientific community as well. In quantum mechanics, for ex-ample, there is a long-standing conjecture that the observer is necessarily an active participant in each experiment. Physicists who subscribe to the so-called Copenhagen interpretation hold that a particle materializes only *after* an obser-vation takes place. Prior to this, the particle exists only as a collection of distinct possibilities, which propagates through spacetime in the form of a probability wave. Since the observer can potentially influence (consciously or unconsciously) which of these possibilities will be realized in the course of the measurement, it is fair to say that the distinction between subjectivity and objectivity becomes blurred on the microscopic level. Indeed, if we define "objective reality" as something that is completely independent of whether or not it is being "probed," then there seems to be no such thing on the quantum level. Why, then, should we automatically dismiss the possibility that this could be the case (at least, in some measure) on the macroscopic level as well?

With that in mind, I think that scientists should not judge anecdotal evidence so harshly, particularly when it is the only kind available. Nor should they automatically rule out historical testimony that describes singular and unprecedented events. Indeed, if quantum physics were related to us as a story, some parts of it would probably sound more fantastic than a fairy tale! The obvious implication here is that compatibility with everyday experience needn't be a necessary condition for rationality (or truth, for that matter). From that standpoint, it would appear that Einstein was not exaggerating when he suggested that:

"Common sense is a collection of prejudices acquired by age 18." [20]

This is not to say that Einstein himself was completely open-minded. His stubborn refusal to accept the inherent uncertainty of quantum mechanics is testimony to the fact that even the most creative human beings find it difficult to embrace the concept of a universe that is full of surprises.

Some Concluding Thoughts

What can we conclude about miracles? It seems that very few phenomena actually deserve such a description. The abundance of human ignorance, superstition and capacity for manipulation suggests that this word must be used with extreme care. Nevertheless, in view of the preceding discussion, it would appear to me that belief in the possibility of miracles has a rational justification, despite the lack of empirical evidence. But this is probably all that a scientist can confidently say on this subject.

> "Scientists have no concern with miracles, for they cannot predict them, bring them about, or draw any conclusions about the future course of nature from them. A miracle is supernatural, and therefore of no scientific interest. ... The scientist ignores the possibility of miracles, just as the lawyer must ignore the possibility of a presidential pardon for his client. ... A pardon is a free action by the President, which cannot be guaranteed by any legal maneuver.
>
> We could not settle whether presidential pardons are possible by looking at the day-to-day business of the courts; rather, we must ask what kind of legal system we live under. We cannot settle whether miracles occur by looking at the ordinary course of nature; we must

ask what kind of universe we live in. This is a philosophical, not a
scientific question." *Richard Purtill*[21]

In the final analysis, we must accept the fact that deep mysteries of faith
cannot be proved, disproved, or even properly discussed in the framework of
science. This type of discourse belongs to the domain of theology, which is
uniquely equipped to deal with such matters. We should also keep in mind that
miracles are first and foremost events of *religious significance*, whose implications
are intimately tied to our belief in God. Interpreting such phenomena requires a
context, which is generally not available to the proverbial "detached observer."

10.2   Can We Investigate the Unknowable?

For more than two centuries secular thinkers have argued that theology lacks
an objective method for distinguishing truth from falsity. Religious claims and
teachings have routinely been portrayed as unverifiable, subjective and altogether irrational. Such criticism is clearly warranted when it comes to superstition or blind adherence to beliefs that contradict established empirical facts.
It is important to recognize, however, that these attitudes usually have very little to do with genuine theology, which encourages a constructive dialogue with
science.

> "Science can purify religion from error and superstition; religion can
> purify science from idolatry and false absurdities. Each can draw the
> other into a wider world, a world in which both can flourish." *Pope
> John Paul II*[22]

With that in mind, in this section we will focus on what I consider to be an
'intelligent' faith, and examine whether its methods of inquiry meet the standards set by science. In addressing these issues, we should, perhaps begin by
considering whether science itself is a purely objective enterprise, with a unified
methodology that extends across all its disciplines. Many contemporary thinkers
have argued against such a simplistic view.

> "The concept of objectivity ... is crude, and tends to suggest false
> dichotomies ... Scientific belief is not the product of us alone or
> the world alone; it is the product of an interaction between our psy-
> chological capacities, our social organization, and the structure of
> the world. The world does not 'stamp' beliefs on us, in science and
> elsewhere." *Peter Godfrey-Smith*[23]

> "What we observe is not nature itself, but nature exposed to our
> method of questioning." *Werner Heisenberg*[24]

> "It is wrong to think that the task of physics is to find out how nature
> *is*. Physics concerns what we can *say* about nature." *Niels Bohr*[25]

Questions have also been raised about the role that formal reasoning and experimental evidence play in the process of scientific discovery. Philosopher Ernan McMullin points out, for example, that certain aspects of physical reality tend to disclose themselves through intuitive insights, which lead to the development of new concepts that may be completely removed from our sensory experience. This process, which he refers to as *retroduction*, is based largely on the creative power of the human imagination, and involves a qualitative "leap" into the realm of the unknown which is not unlike the one that theologians make. The importance of retroductive reasoning is perhaps most apparent in the domain of cosmology, where we routinely make extrapolations from our own limited experience to the universe as a whole (which obviously eludes our methods of observation).

> "Cosmology of its very nature demands extrapolation, often quite daring extrapolation. Because its objects are distant and unfamiliar, it has always had to rely on indirect and precarious modes of reasoning." *Ernan McMullin* [26]

It is true, of course, that we have no logical grounds to expect that anything other than standard forms of scientific reasoning (such as induction and deduction) can lead to verifiable knowledge about nature. And yet, we know that this is not always the case. Indeed, the history of science abounds with examples of discoveries that were based on purely theoretical speculations, which had no basis in observable evidence. Nobel Prize winning physicist Wolfgang Pauli provides a striking illustration of this point.

> "The bridge, leading from initially unordered data of experience to ideas, consists in certain primeval images pre-existing in the soul. ... These primeval images should not be located in consciousness or related to specific rationally formulizable ideas. It is a question, rather, of forms belonging to the unconscious region of the human soul, images of powerful emotional content, which are not thought but beheld, as it were, pictorially. The delight one feels, on becoming aware of a new piece of knowledge, arises from the way such pre-existing images fall into congruence with the behavior of the external objects. ... One should never declare that theses laid down by rational formulation are the only possible presuppositions of human reason." [27]

Such a view of science is far more complex than the traditional one, and suggests that social, psychological and aesthetic factors are of considerable importance. If we agree, then, that science is not just about mathematical formulae and measurable quantities, and involves values, imaginative schemes and human judgement as well, it becomes far more difficult to claim its superiority over theology. With that in mind, we now proceed to consider several specific areas where the empiricist and religious outlooks can be meaningfully compared.

The Nature of Religious Knowledge

In evaluating the methods of theology and science, it must first be recognized that their subject matter is fundamentally different. In the case of the natural sciences, we have considerable control over the object of our study – experiments can generally be repeated with virtually identical outcomes (chaos being a notable exception in this respect). The questions that science asks tend to be very precise, and often allow us to disregard a wide array of "secondary" influences. As a result, we can form relatively simple models that accurately describe many natural phenomena.

As we move to social sciences, our level of control diminishes dramatically and the methods necessarily change. In psychology, for example, it is not possible to establish what is in the mind of the subject unless she chooses to tell us herself. For that reason, the ability to gain someone's trust becomes just as important as experimental techniques.

History is also relatively imprecise, given that it deals with complex questions where causal relationships are difficult to establish. Unlike classical mechanics or electric circuit theory, in this case it is hard to distinguish "secondary" effects from those that significantly influence the course of events. It could be plausibly argued, for example, that the mental and physical state of each soldier might be crucial for the outcome of a battle. It might also be relevant to consider what the army had for breakfast, whether it rained during the night, and a whole host of other seemingly trivial details.

Along these lines, philosopher Isaiah Berlin eloquently described how the research method of history differs from the method of mathematics and physics.

> "Whereas in a developed natural science we consider it more rational to put our confidence in general propositions and laws than in specific phenomena, this rule does not seem to operate successfully in history. ... Addiction to theory – being doctrinaire – is a term of abuse when applied to historians; but it is not an insult if applied to a natural scientist. ...
>
> In a developed work of natural science ... the links between the propositions are, or should be, logically obvious. ... Even if such symbols of inference as 'because', or 'therefore', were omitted, a piece of reasoning in mathematics or physics ... should be able to exhibit its inner logical structure by the sheer meaning and order of its component propositions. ... This is very far from being the case in even the best, most convincing, most rigorously argued works of history. ...
>
> The business of a science is to concentrate on similarities, not differences, to be general, to omit whatever is not relevant to answering the severely delimited questions that it sets itself to ask. ... Historians are interested at least as much in the opposite: in that which differentiates one thing, pattern of experience, individual or collective, from another. When such historians attempt to account for and

explain, say, the French Revolution, the last thing that they seek to
do is to concentrate only on those characteristics which the French
Revolution has in common with other revolutions, ... to formulate
a law from which something about the pattern of all revolutions ...
could in principle be reliably inferred. ...

The immediate purpose of narrative historians ... is to paint a
portrait of a situation or a process, which, like all portraits, seeks
to capture the unique patterns and peculiar characteristics of its
particular subject; not to be an X-ray which eliminates all but what
a great many subjects have in common. ... Historical explanation
is to a large degree arrangement of the discovered facts in patterns
which satisfy us because they accord with life." [28]

The above discussion suggests that the mode of investigation which we employ in a particular field depends to a large extent on the subject that it is
concerned with. A physicist deals with inanimate matter, and can effectively
utilize experimental observations as well as mathematical generalizations. The
subject matter of history is far more elusive, and provides considerable room for
different (and even conflicting) interpretations. By this standard, theology is on
the opposite pole from science, in the sense that it has no control whatsoever over
its subject. Religious experiences (which provide much of the empirical evidence
for theological claims) tend to be personal, and cannot be replicated at will. As
a result, it is virtually impossible to achieve the level of "objective consensus"
that characterizes the natural and social sciences. This means (among other
things) that theology must derive its conclusions using a somewhat different set
of criteria, which are unique to its subject of inquiry.

There is no reason to view this methodological diversity in a negative light.
Even within science itself, each discipline has its own language and epistemological standards. In quantum mechanics, for example, we are forced to use
probabilistic interpretations, since the microscopic world simply doesn't lend itself to a precise description. In electric circuit theory the situation is quite the
opposite - differential equations tend to work very well, and allow us to make accurate predictions. These two disciplines approach their subject matters in very
different ways, but it would be incorrect to claim that one of them is "superior"
to the other.

In this context, we should also mention that criteria for a 'good' explanation
vary widely across science. [29] In some instances, a simple description of the mechanism is considered to be sufficient (as in the case of DNA replication). In other
situations, much more information needs to be provided. Explaining the motion
of a pendulum, for example, requires a system of differential equations and an
exact knowledge of the initial conditions. All of these illustrations underscore
the fact that truth comes in different forms, and that ways for discovering it
vary accordingly. As a result, we can legitimately argue that it is inappropriate
to represent scientific knowledge as superior, or to view its method as the only
acceptable form of inquiry.

Empirical Aspects of Religion

If we agree that religious knowledge eludes the methods of science, then it is necessary to provide some rational and systematic alternative for discerning truths about God. What does theology have to offer in that respect? To begin with, it is important to recognize that theology is more like a road map than like a scientific theory:

> "To suppose that people can be saved by studying and giving assent to formulae is like supposing that one can get to Timbuktu by poring over a map of Africa. Maps are symbols, and even the best of them are inaccurate. But they are indispensable tools to reach a given destination." *Aldous Huxley* [30]

We should also remember that theological knowledge is a means, not an end in its own right:

> "Some men love knowledge and discernment as the best and the most excellent of things. Behold, then the knowledge and discernment come to be loved more than that which is discerned." *Theologia Germanica* [31]

Mystics of all traditions agree that only direct experience can provide a real knowledge of God. St. Thomas Aquinas acknowledged this explicitly towards the end of his life, when after a contemplative experience he refused to continue his theological work. In his own words:

> "Everything I had written up to now is as a mere straw compared to the immediate knowledge of my experience." [32]

Such testimony suggests that theology should be primarily concerned with identifying a systematic approach that can lead to a unitive experience of Divine reality.

It would be naïve, of course, to look for an algorithm for achieving religious enlightenment (although I have seen a few books that claim to do so). Most traditions take a more modest approach, and agree that meditative prayer is an appropriate method for enhancing one's spiritual state. The fact that this process does not yield instantaneous results does not diminish its credibility – in that respect, it is no different from science. Indeed, a scientist who focuses on a problem often spends an extraordinary amount of time thinking about it. It may take years before anything happens, and if it does, it does so in a most unexpected way and place. This is how the great French mathematician Henri Poincaré recalled his experience of enlightenment:

> "Having reached Coutances, we entered an omnibus to go to some place or other. At the moment when I put my foot on the step the idea came to me, without anything in my former thoughts seeming to have paved the way for it, that the transformations I had used to define the Fuchsian functions were identical with those of non-Euclidean geometry." *Henri Poincaré* [33]

It is fair to say that creative breakthroughs like Poincaré's are rooted in an overall attitude toward science. Thinking about a problem on a regular basis without being affected by the day to day results is often the only way to set the conditions for a real scientific revelation. The same is true of meditation and prayer – it is simply practiced until the conditions are right for 'something' to happen. We have no conscious control of if, when, and how this will take place.

An important point for both scientific and religious practices is to avoid attachment to outcomes, and to take pleasure in the process itself.

> "St. Ignatius Loyola was once asked what his feelings would be if the Pope were to suppress the Company of Jesus. 'A quarter of an hour of prayer', he answered, 'and I should think no more about it.'[34]

Such 'holy indifference' may be difficult to achieve, but the road that leads to it is relatively well defined. In principle, this road is as open to scientists and mathematicians as it is to great religious figures.

> "If we concentrate our attention on trying to solve a problem of geometry, and if at the end of an hour we are no nearer to doing so than at the beginning, we have nevertheless been making progress each minute of that hour in another more mysterious dimension. Without knowing or feeling it, this apparently barren effort has brought more light into the soul. The result will one day be discovered in prayer." *Simone Weil*[35]

The Explanatory Power of Religion

In justifying the apparent superiority of the scientific method, many secular thinkers have pointed to the fact that theology lacks explanatory power.

> "An explanation builds a conceptual bridge from the known to the unknown, linking the unexplained to the context of one's knowledge." *George Smith*[36]

Smith goes on to say that logic does not allow us to explain the unknown with references to the unknowable, as some do when they explain miracles and mysteries by reference to God.

In evaluating such claims it is useful to recognize that the concept of "scientific explanation" is rather broad. It is often said that a phenomenon has been "explained" if we can predict its course under different circumstances. In classical mechanics, for example, we can fairly easily identify what causes the system to behave in a particular way – all the necessary information is contained in the differential equations and the initial conditions. When it comes to more complex phenomena, however, it becomes impossible to isolate all the relevant parameters and predict exactly what will happen (the throwing of dice is a typical illustration). We tend to refer to such processes as *random*, but this doesn't automatically mean that they have no explanation. If, for example, we decide to flip a coin 10,000 times, we really cannot say anything about the outcome of the first toss, but we do know that the number of recorded heads and tails will be approximately equal at the end of the experiment. Since there is some sort of high – level regularity in this case, a "statistical" explanation is entirely appropriate.

There are, however, certain types of randomness that defy even statistical explanation. To better understand what this means, try closing your eyes and landing a pen ten times on a piece of paper. Then ask yourself whether the resulting arrangement of points can be "explained" in any meaningful way. Before answering this question, it is helpful to recall that you can *always* come up with a function that passes through all of these points (doing that systematically would require the construction of a 9-th order interpolation polynomial). There is, in other words, a well-defined mathematical model for this apparently random phenomenon. The problem, however, lies in the fact that the model is just as complicated as the data that it is supposed to explain! Indeed, a 9-th order polynomial is uniquely defined by 10 coefficients, which is a set no smaller than the one we started from. This model also fails to provide any useful information regarding the distribution of further points that you might want to add. If you were now to mark another 9,990 points, your old polynomial would be of no use and you would have to construct another one of order 9,999 in order to fit the data. In cases like this, we say that there is no way to *compress* the information content of the data, and the whole concept of an "explanation" becomes rather meaningless.

The notion of a scientific "explanation" is also related to the idea of cause and effect. If we say that event A causes event B, this usually implies that all recorded occurrences of A were followed by B. The problem with this interpretation is that it implicitly assumes that all future experiments involving event A will continue to yield B as a part of the outcome. It is well known, however, that this needn't always be the case. Chaotic systems like the one described in Fig. 10.1 provide a perfect counterexample, since they allow for sudden and dramatic disruptions in their otherwise regular dynamic behavior.

The concept of causality is particularly problematic in quantum mechanics, where measurement results cannot be adequately "explained." Schrödinger's equation allows us to causally relate the *probabilities* of different outcomes, but it tells us nothing at all about why one particular value was measured and not some other. In quantum mechanics, there is actually no point in asking *why* something happened - the only meaningful question is *what* happened. In view of that, it is not unreasonable to suggest that our conventional concept of causality is perhaps no more than a useful illusion.

"The law of universal causation is an attempt to bolster up our belief that what has happened before will happen again, which is no better founded than the horse's belief that you will take the turning you usually take." *Bertrand Russell* [37]

"Modern man has used cause and effect as ancient man used the gods to give order to the universe. This is not because it was the truest system, but because it was the most convenient." *Henri Poincaré* [38]

In view of the difficulties associated with prediction and causality, scientists often tend to relax the criteria for what constitutes an adequate 'explanation'. Indeed, in areas of physics where empirical evidence is scarce (such as string theory), the primary criterion is whether the proposed explanations render the world more *intelligible* or not. This view finds support among contemporary philosophers of science as well. Philip Kitcher, for example, has argued that providing a scheme (or theory) that unifies a range of phenomena can count as a valid explanation in science, although the theory itself may provide no specific predictions, or may be unsuitable for experimental testing.[39] With that in mind, the proper question to ask of theology is whether belief in God helps us make sense of our existence, and our place in the universe.

In evaluating the rationality of religious 'explanations', it is important to recognize that they make sense only at the most fundamental level of knowledge. They apply to questions that exceed the scope of scientific reasoning, such as the existence and meaning of the universe, or the origins of physical laws. The fact that science cannot explain its own axioms provides room for theological speculation, in which intelligibility and internal consistency are legitimate criteria for rationality.

> "As we reach the boundary of the realm where the writ of science does not run, experimental verification or falsification is no longer possible but we can still reasonably require that our account should be self-consistent. A minimum criterion for any rational explanation is therefore that of consistency." *David Bartholomew* [40]

One could argue, of course, that we accept string theory and the Big Bang model precisely because they have *some* predictive power, and are at least potentially verifiable by experiments. This is clearly not the case with theological statements (as long as the various prophecies that permeate religious literature are not taken literally). We should note, however, that evolution theory suffers from the same problem – it has no predictive power, and yet we consider it to be a perfectly legitimate scientific model. The fact of the matter is that Darwin's theory tells us nothing at all about where the process of evolution will take us in the future. It simply organizes existing evidence into a coherent whole.

In their critique of theology, secular thinkers also emphasize the fact that scientific experiments are *reproducible*, and that the outcomes can be verified by anyone with sufficient technical means and expertise. This is a feature that seems to be altogether absent from religious discourse. In responding to such objections, it should be noted that meditative states *are* reproducible when following a certain methodology. The fact that there is no explanation for the effectiveness of these techniques is by no means unusual – this type of situation is quite common in experimental sciences (biology, plasma physics and areas of nuclear physics are typical examples).

A persistent skeptic could expand this argument, and observe that scientific experiments generally yield *consistent results*, while reports of mystical experiences vary greatly across (and even within) different traditions. It seems to me that there are two possible ways to counter such claims. In the first place, it is not true that all scientific experiments have predictable outcomes. It is well known, for example, that we can never anticipate what a quantum measurement will produce, even if the external conditions are known precisely. The most we can say in such cases is that a certain statistical pattern emerges when a large number of identical systems are examined. The famous physicist Richard Feynman described this property in the following way:

> "A philosopher once said: "It is necessary for the very existence of science that the same conditions always produce the same results." Well, they don't!" *Richard Feynman* [41]

A second way to defend the validity of mystical practices is to note that uniformity of outcomes is *not* a necessary condition for accepting empirical evidence – it is only a *sufficient* one. An analogy from chaos theory may help to clarify this point. In chaotic systems there is a wide variety of unpredictable outcomes, with the attractor as the principal unifying empirical characteristic. However, this apparent diversity clearly doesn't imply the lack of a common underlying principle (in this case, the equation), or a method of investigation (nonlinear system theory). With that in mind, there appears to be nothing a priori illogical in the fact that mystics report very different experiences, and we have no reason to suppose that theology cannot study their claims in a systematic manner.

The Acceptance of Religious Authority

Theological claims are often justified by references to authority (the Bible, major religious figures of the past or the canons of the Church). This is fundamentally opposed to the scientific method, which assumes that theories can be independently verified by all who possess the necessary expertise. It is therefore worth asking whether reference to authority constitutes an intellectually acceptable mode of inquiry.

In responding to this question, we must once again take into account the unique subject matter of theology. Huston Smith pointed out that even the American constitution, which is a relatively simple document, requires a Supreme Court to interpret it.[42] How, then, can one expect to understand something as complex as the Bible without some help from authorities on the subject? Biblical writers were primarily concerned with describing the interaction of the human and the Divine, which requires the use of *symbolic* language. It is therefore reasonable to expect that the true meaning of this text cannot be grasped without considerable theological knowledge.

A somewhat different way to justify religious authority stems from the need for *models*. Since spirituality eludes theoretical or experimental explanations, one must rely (at least initially) on the testimony of those who have had such experiences. The Buddha compared this process to the training of a wild elephant. According to him, the best way to do this is to yoke it to one that is already trained. Through contact, the wild animal recognizes that being an elephant is actually compatible with being yoked, and that this objective is attainable.[43]

The commandment of Jesus that we should "Go and do likewise" is another example of a religious model. It is not a strict blueprint for spiritual enlightenment, nor is it confined to a particular time and place. Nevertheless, it is a legitimate (and very robust) model for Christian life.

> "The basic commandment that Jesus gives at the end of the Good Samaritan story invites Christians to think analogically: "Go and do likewise" (Luke 10:37). The mandate is not "Go and do exactly the same" as the Samaritan. It is decidedly not "Go and do whatever you want." The term 'likewise' implies that Christians should be faithful to the story of Jesus yet creative in applying it to their context." *William Spohn* [44]

Even if one has had a personal religious experience, a certain amount of guidance is necessary in order to properly interpret it. Left to our own devices, we would have a hard time explaining what has happened, and we might easily be tempted to dismiss the entire event as an illusion. This would actually be a perfectly natural reaction, given that knowledge typically requires external validation before it is properly assimilated. Along these lines, sociologists of knowledge have argued that what we find believable generally conforms to what is acceptable to society as a whole. This, in turn, depends to a large extent on the views and opinions promoted by figures of authority.

Based on the above discussion, it would seem that there is nothing intrinsically irrational or unscientific in authoritative teaching. In areas that do not lend themselves to mathematics or direct experimental verification, this is actually a perfectly natural starting point. To some extent, this is true even in the sciences, where an implicit trust in the authority of the teacher necessarily precedes direct verification of what is taught. We might add in this context that we have no reason to suppose that there is no verification at the end of the religious journey, although this may take a lifetime (or perhaps longer).

Can Theology Incorporate New Scientific Knowledge?

Many secular critics point to the rigid nature of religious dogma, and claim that theology shows no willingness to revise its teachings in light of new scientific knowledge. I would argue that this is not a valid objection. It is true that the articles of faith are in some sense like axioms – they are fixed, and needn't be justified. The teachings of religion, however, are more like theorems, which are derived from the fundamental principles with a certain logic and method (something akin to rules of inference). Both religious and secular thinkers have recognized this structural similarity between mathematics and theology.

> "The form of Newton's Principia, in spite of its admittedly empirical material, is entirely dominated by Euclid. Theology, in its exact scholastic forms, takes its style from the same source. Personal religion is derived from ecstasy, theology from mathematics." *Bertrand Russell* [45]

"As the other sciences do not argue in proof of their principles, but argue from their principles to demonstrate other truths in these sciences, so this doctrine does not argue in proof of its principles, which are the articles of faith, but from them it goes on to prove something else." *St. Thomas Aquinas* [46]

We have no a priori reason to suppose that the "theorems" of theology cannot change over time, given the availability of new knowledge. There is, in fact, significant historical evidence in favor of the view expressed above - the tradition of revising religious teaching goes back to the early Middle Ages. Much of the Church doctrine as we know it today was formulated at that time, long after the death of Christ. There is no doubt that these teachings were based on the events reported in the New Testament, but the interpretations were new and original. This process is quite consistent with the methods of modern science, where we often discover theoretical meaning long after the initial observations occur.

Another illustration of doctrinal flexibility can be found in the work of Aquinas, who incorporated much of Aristotle's thought into his theological writings (this was the best available science in the 13th century). More recently, Pope John Paul II openly acknowledged the legitimacy of evolution theory, while pointing out some of its limitations as well.

"New knowledge has led to the recognition of the theory of evolution as more than a hypothesis. .... The sciences of observation describe and measure the multiple manifestations of life with increasing precision and correlate them with the time line. The moment of transition to the spiritual cannot be the object of this kind of observation." *Pope John Paul II* [47]

In light of the above examples, I think it would be inaccurate to claim that theology is (or ever has been) simply a collection of invariant ideas and teachings. Instead, it may be more appropriate to describe it in terms of *paradigms*, which occasionally change as new knowledge becomes available. [48] In that respect theology is not too different from science, which has its own periodic paradigm shifts (usually triggered by anomalies which existing theories cannot explain). This process may be much faster in the case of science, but the transition is nevertheless a painful one, with much resistance from conservative forces.

"A new scientific truth does not triumph by convincing its opponents and making them see the light, but rather because its opponents eventually die, and a new generation grows up that is familiar with it." *Max Planck* [49]

Planck's statement clearly undermines the myth of the "perfectly objective" scientist, who is willing to invalidate his previous work in the name of "truth." This is an important and useful insight, since such stereotypes often cause confusion and create unrealistic expectations. The fact of the matter is that scientists are fallible, and are just as prone to biases and conservativism as any other human being.

"The transfer of allegiance from paradigm to paradigm is a conversion experience that cannot be forced. Lifelong resistance, particularly from those whose productive careers have committed them to an older tradition of normal science, is not a violation of scientific standards but an index to the nature of scientific research itself. The source of resistance is the assurance that the older paradigm will ultimately solve all its problems, that nature can be shoved into the box that the paradigm provides. Inevitably, at times of revolution, that assurance seems stubborn and pigheaded as indeed it sometimes becomes." *Thomas Kuhn* [50]

When comparing science and theology along these lines, it is important to stress that there is nothing inherently wrong in believing in an existing paradigm, although we may be aware of its potential limitations and incompleteness. As Kuhn pointed out, science is characterized by prolonged periods of "normal" development, during which a particular set of assumptions and theories remains largely unquestioned. Such periods create the necessary conditions for discovering anomalies, and provide the appropriate context for interpreting them (keep in mind that something can be called an "anomaly" only if we already have strong expectations about what reality ought to look like). Without this kind of implicit belief, our capacity for creating new knowledge would be severely limited, and science would develop at a very different pace (if at all). Why, then, should we require a different and more demanding standard for theology?

10.3   Can Religion Be Reconciled with Evolution?

Evolution theory has been an area of conflict between science and religion for over a century. On one pole of this debate are the religious 'literalists', who claim that the Bible provides an exact and indisputable account of creation. On the other extreme are the scientific materialists, who see the world and everything in it as a result of the impersonal laws and forces of nature. These two positions are so fundamentally opposed to each other that it makes little sense to try to reconcile their claims. There are, however, more moderate opinions on both sides of this issue, which allow for a constructive dialogue. In the following, we will explore this middle ground, and consider whether the views of theology and science regarding evolution can eventually be harmonized in some measure.

Does Evolution Have a Direction?

Perhaps the most fundamental question related to evolution is whether or not it has a 'direction.' Many scientists have argued that it does not.

"Natural selection is a blind, unconscious, automatic process that has no purpose in mind." *Richard Dawkins* [51]

Biologist Stephen Jay Gould makes the same point and questions the view
that evolution is "progressive," leading toward advanced forms such as consciousness.

> "Darwin maintained that evolution has no direction; it does not lead
> inevitably to higher things. ... The degeneracy of a parasite is as
> perfect as the gait of a gazelle." *Stephen J. Gould*[52]

From a theological standpoint, views such as those of Dawkins and Gould
pose a serious challenge, since they appear to be at odds with the notion of God's
purposeful design. Indeed, it is difficult to imagine how random processes like
mutation or natural selection can be associated with any kind of preconceived
order or structure. With that in mind, in this section we will explore what system
theory and information science have to say regarding the possible coexistence of
uncertainty and global organization.

We begin with a hypothesis that is widely accepted in the scientific community, and is rooted in the results of modern physics and theoretical biology. This
hypothesis states that ever since the Big Bang there has been a trend in the
universe toward increased complexity, with human consciousness as its highest
recorded level to date.

> "There exists alongside the entropy arrow another arrow of time,
> equally fundamental and no less subtle in nature. Its origin lies
> shrouded in mystery, but its presence is undeniable. I refer to the
> fact that the universe is progressing - through the steady growth of
> structure, organization and complexity - to ever more developed and
> elaborate states of matter and energy." *Paul Davies*[53]

> "To assert blithely that evolution proceeds by purely chance events is
> much less than a precise description. ... Though its actual course is
> indeterminate, its general course towards complexity, self-organization,
> and even the emergence of self-replicating molecules and systems ...
> can be interpreted as inevitable in the universe in which we live."
> *William Stoeger*[54]

In interpreting views such as those articulated above, it is important to
exercise a certain amount of caution. It would be a mistake to assume, for
example, that the emergence of complex structures in nature (including life
itself) automatically implies that evolution follows a definite "plan." Indeed,
one could legitimately argue that although there is a certain discernible pattern
in the way the universe has developed, this property needn't entail the existence
of a specific preconceived goal, or necessitate a uniquely defined process. In other
words, it is entirely reasonable to think of evolution in terms of a *multiplicity* of
directionalities, which coexist and possibly interact on different levels.

It goes without saying, of course, that any increase in complexity must occur
in accordance with the laws of nature, which limit the range of possible forms
that can arise at any given stage of the process. We should add, however, that
these laws appear to be quite "liberal," in the sense that they allow for an

almost unlimited variety of configurations. In view of that, one could perhaps compare the universe to a giant laboratory, in which countless "experiments" are being conducted, both simultaneously and sequentially. A small number of these experiments succeed in initiating a chain of further developments, but most do not. It would be natural to assume in this context that the "surviving" forms are the ones that provide some sort of competitive advantage. That is, after all, how most people perceive the process of natural selection. We should point out, however, that such interpretations tend to be overly simplistic, because they disregard the fact that the external environment constantly changes. These changes are often random, and their magnitude and duration are essentially unpredictable. Under such circumstances, it would be difficult to define what we mean by "competitive advantage," since variations that are beneficial in one type of environment may be detrimental in a slightly different one.

The existence of many "dead ends" and the apparent randomness with which the surviving forms are selected are well known characteristics of evolution, which were recognized by Darwin himself. It is important to keep in mind, however, that this does not imply that evolution is necessarily devoid of direction, or that we must automatically rule out the possibility of a Divine purpose. On the contrary, it is not unreasonable to expect that a God who loves his creation would allow it the freedom to explore a broad range of possibilities. From this perspective, the randomness that characterizes genetic variations and natural selection could be viewed as a "mechanism" that opens up new developmental possibilities, and leads to structures and forms which were previously infeasible. If the number of such "experiments" is very large, it is entirely plausible to expect that the overall process will ultimately produce the right mix of freedom and order that is necessary for the emergence of life.

The following simple example illustrates how a combination of random and orderly features can gradually give rise to elaborate structures and behavioral patterns that are comparable to those encountered in biological systems.

Example 10.1 (Sierpinski Triangles). Consider an arbitrary triangle in a plane, whose vertices are given by coordinates (x_1, y_1), (x_2, y_2) and (x_3, y_3). Let us pick a starting point (x_0, y_0) anywhere within the triangle, and form the following iterative sequence:

$$x(k+1) = 0.5\,[x(k) + x_R]$$
$$y(k+1) = 0.5\,[y(k) + y_R] \tag{10.1}$$

where the pair (x_R, y_R) is *randomly* chosen in each step from the set of vertices (x_1, y_1), (x_2, y_2) and (x_3, y_3). Amazingly, this sequence always converges to the attractor shown in Fig. 10.2, which is known as a Sierpinski triangle.

The attractor shown above has another surprising feature, which is related to the coefficients of the binomial expansion (see Section 7.3). Namely, if we were to set all even entries in Pascal's triangle to 0 and all odd ones to 1, we would obtain exactly the same geometric form as in Fig. 10.2 (assuming that the ones are represented as points, and zeros as empty spaces).

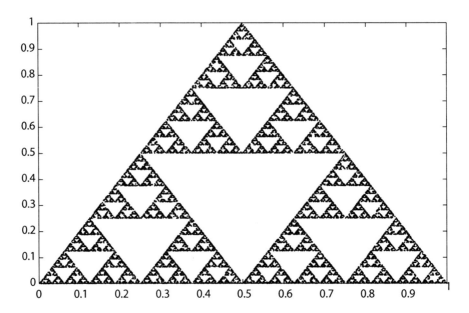

Fig. 10.2. The fractal structure associated with equation (10.1).

In order to draw an analogy between equation (10.1) and biological systems, let us now imagine that the arbitrarily chosen points x_R and y_R represent "random mutations" and that (10.1) is a lawful way of transmitting them to the next "generation." Viewed in this light, it would appear that the outcome of the selection process is not random at all (although this is by no means apparent from local observations). It is true that we must execute a large number of iterations before any sort of complex structure becomes noticeable, but this is the case with biological evolution as well. In both instances, a high-level regularity emerges over time, whose subtle details may be visible only from "outside" the system.

Example 10.1 underscores the fact that a combination of randomness and simple rules can spontaneously produce intricate and orderly global structures over prolonged periods of time. This pattern of behavior is commonly encountered in complex systems, and is associated with the phenomenon of *self-organization*. As noted in Chapter 2, self-organization plays a central role in the development of living organisms, and is characterized by the fact that local interactions give rise to global structures and increased complexity *without* any "master plan."

On first glance, such a "decentralized" model might appear to be inappropriate for the study of living organisms, since molecular biology suggests that the development of cells, tissues and all higher level functions *does* follow a sort of compressed "blueprint" (the one stored in the genome). This is clearly a valid point, since the information contained in the DNA of any organism initiates the developmental process, and constrains the possible outcomes. However, it is also true that the details ultimately depend on many other factors. Stuart Kauffman's work on random Boolean networks has shown, for example, that the nature of the interactions among genes plays a crucial role in determining certain important characteristics of living organisms (see Section 2.3 for more details).

Recent work in the emerging field of epigenetics provides further evidence that DNA alone does not fully determine how individual genes operate. Much of the information that shapes this process apparently resides in a collection of chemical "switches" which are distributed along the double helix structure of the DNA. It is these switches (which are part of what is known as the *epigenome*) that ultimately decide whether a particular gene will be activated or deactivated.

Experiments by Randy Jirtle and his collaborators at Duke University have shown that the epigenome is sensitive to our physical environment, and is also affected by our behavioral and dietary patterns. Among other things, they found that a gene which increases the risk of cancer and diabetes in mice can be suppressed simply by changing the mother's diet (without altering the DNA sequence in any way). Even more interestingly, it now appears that changes in the epigenome due to external influences can be propagated across several generations. Recent studies have shown that such changes are not limited to the early stages of fetal development, and can occur throughout an individual's lifetime.

> "Epigenetics is proving we have some responsibility for the integrity of our genome. ... Before, genes predetermined outcomes. Now everything we do – everything we eat or smoke – can affect our gene expression and that of future generations. Epigenetics introduces the concept of free will into our idea of genetics." *Randy Jirtle* [55]

What can we conclude from all this? Based on the above discussion, it would appear that chance is actually conducive to increased complexity, and that the randomness of natural selection does not automatically rule out the possibility of a Divine purpose. When making such a claim, it is important to recognize that there is a subtle but crucial difference between terms "Divine purpose" and "Divine design." Theologian Robert John Russell explains this distinction in the following way:

> "The potentialities of the universe and its individual entities are not in the form of a blueprint of the future, so it is misleading to speak of divine design. The term 'design' has connotations of a preconceived detailed plan. ... The term 'purpose' is better as it does not carry this connotation. Nothing is completely determined. The future is open ended." *Robert J. Russell* [56]

Russell's description of evolution is clearly consistent with the notion of a God who does not "micromanage" his creation. It is entirely reasonable to assume that such a God would act by providing a range of *possibilities* and *constraints* (in the form of natural laws), with the knowledge that the system will ultimately self-organize in a manner that conforms to His purposes. On this view, God's role in evolution is far more subtle than what Biblical literalists claim, and involves a particular choice of natural laws which are conducive to self-organization and increasing complexity. Interestingly, it was Darwin himself who opened the door to such speculations:

> "I am inclined to look at everything as resulting from designed laws, with the details, whether good or bad, left to the working out of what we may call chance." [57]

Are Humans "Special"?

Even if we agree that evolution has some form of directionality, this does not automatically imply that humans are entitled to a special status in the animal kingdom. In fact, the level of similarity in the genetic structures of primates is such that the opposite conclusion seems more likely.

> "Modern science is rapidly removing every excuse ... that we are much different from our closest primate relatives." *Steven J. Gould* [58]

The suggestion that there is no qualitative difference between humans and apes obviously conflicts with the Christian view that man was created in the image of God. With that in mind, in the following we will examine whether science allows for some sort of compromise between these two positions.

Nonlinearity and Emergent Phenomena

We begin by questioning the assumption that small genetic variations necessarily result in small morphological and behavioral differences. Although this would be a perfectly reasonable conclusion for a linear system, it is quite possible that evolution is actually a *nonlinear* process (like most other complex natural phenomena). It is well known that such processes allow for major qualitative changes in the system even when the perturbation is very small (see Chapter 2 for a more detailed discussion of this property).

In support of this conjecture, we might point to a number of highly nonlinear chemical reactions which are essential for sustaining all living organisms. A typical example is the process of *autocatalysis*, which is responsible for regulating metabolic activity on the cellular level. [59] Autocatalysis is usually encountered in systems that are not isolated, and are exposed to a constant influx of matter and energy from the external environment. Such systems tend to be in a state of self-organized criticality, which is closely associated with chaos.

Although the relationship between chaos and self-organized criticality was already discussed in Chapter 2, it might be helpful to supplement that description with an illustrative example. In order to do that, imagine a small pile of sand which is placed on a flat surface. If we were to slowly add grains to the pile, it would gradually grow until it reached a critical height. Supplying a single new grain at that point would create a small avalanche, which would temporarily improve the stability of the system. However, adding more grains of sand would soon bring the system back to the critical level, and the process could repeat (although never in exactly the same way). Such a system would perpetually fluctuate around the critical point, which represents a state of elevated complexity. It is interesting to note that this simple model is actually quite general, and applies to a variety of phenomena ranging from species extinction to stock market activity.[60] This suggests that complex systems apparently "prefer" to function at the edge of chaos, where creative possibilities are virtually unlimited.

If we agree, then, that evolution is probably a nonlinear process with a tendency toward self-organized criticality, then there is a real possibility for emergent phenomena, which introduce qualitative changes in the system. With that in mind, it is not unreasonable to speculate that the appearance of human beings could have marked a new and unprecedented event in natural history. What we know about nonlinear systems suggests that there is no need for "intermediate" stages between two levels of complexity. Indeed, chaos theory teaches us that when a system is close to its bifurcation point, emergent phenomena can occur suddenly, often as a result of minute perturbations in the environment.

This conjecture is consistent with the ideas of biologists Stephen Jay Gould and Niles Eldredge,[61] who argued in the 1970s that macroscopic evolutionary patterns are often characterized by lengthy periods of relative stagnation, punctuated by occasional bursts of rapid change. These periods of change tend to be unpredictable both in terms of when they will occur, and in terms of what will actually take place. As such, they bear a considerable resemblance to the phenomenon of intermittency in nonlinear systems, which typically occurs on the borderline of chaos (see Figure 10.1 for a visual illustration).

A typical example of an "intermittent" biological process is the development of the human brain. Some three million years ago, the hominid brain weighed approximately 900 grams. Over a period of one million years (which is considered brief by evolutionary standards), the brain mass increased to 1.4 kilograms, and then the growth apparently stopped. It is interesting to note that many of the attributes that distinguish humans from animals (such as speech, tool making and foresight) emerged immediately after the increase in brain size was completed.

Human Consciousness

Those who believe that our species deserves a special place in the hierarchy of nature often base their arguments on the uniqueness of human consciousness. This point of view usually entails the assumption that there is a qualitative difference between the human mind and known forms of matter. It is important to recognize, however, that such an outlook is by no means universal. Francis Crick (one of the pioneers of modern genetics) offers a very different interpretation:

> "You, your joys and sorrows, your memories and your ambitions, your sense of identity and free will, are in fact no more than the behavior of a vast assembly of nerve cells and their associated molecules."
> *Francis Crick*[62]

In order to put statements like this into proper perspective, we first need to consider some possible objections to their central premise. One can legitimately question, for example, the assumption that the behavior of the "whole" can always be reduced to the dynamics of the constituent parts. Perhaps the most striking counter-example in this context are emergent phenomena, which are closely related to nonlinear systems and complexity. In this case, the system components interact in a way that gives rise to qualitatively new properties which are not exhibited by the components themselves. The same can be said of quantum entanglement, and many processes that occur in the domain of biology.

We should also note in this context that the existence of a "whole" often constrains the behavior of the parts that make it up. It is well known, for example, that extraordinarily complex systems such as human societies cannot be simply reduced to the behavior of their constituents. In this case the influence works in *both* directions. Indeed, it is obvious that the existence of laws, social norms and economic relations results in very real limitations on the behavior of each individual.

A fervent reductionist would probably respond to such arguments with the observation that complexity is merely a consequence of deterministic laws and random accidents. As a result, what we refer to as the "mind" is really no more than an epiphenomenon, whose origins (and explanation) are purely physical. If this were true, however, we would have no good reason to believe that our pronouncements reflect objective reality in any meaningful way. Indeed, why would a collection of interacting neurons be expected to make verifiable statements about itself and its environment?

> "If utterances of minds are the result of the interplay of blind determinism and blind chance then it is not clear what relationship these utterances bear to the real world they purport to describe." *David Bartholomew* [63]

> "[W]ith me the horrid doubt always arises whether the convictions of man's mind, which has been developed from the mind of the lower animals, are of any value or at all trustworthy." *Charles Darwin* [64]

In further evaluating the merits of reductionist theories, we should observe that there is a fundamental difference between saying that mind is *related* to matter, and that mind is *reducible* to matter. If we consider that there are very few atoms remaining in our bodies from three years ago, the conjecture that matter alone constitutes our 'consciousness' is not very convincing. If that were the case, the whole concept of 'identity' would be questionable, since it requires a certain degree of temporal continuity. It seems more plausible to view the mind as the general pattern in which this matter is organized. Such a pattern would be an information bearing entity, which would use the brain much like a chess player uses the board and the pieces. Based on this analogy, it seems perfectly rational to speculate that consciousness could exist without matter (just like a chess game continues to exist as a form of information even when the pieces are removed). Such a conjecture finds support in the results of modern physics, which suggest that information transcends matter and can exist independently:

> "What is the universe made of? A growing number of scientists suspect that information plays a fundamental role in answering this

question. Some even go as far as to suggest that information-based concepts may eventually fuse with or replace traditional notions such as particles, fields and forces. The universe may literally be made up of information, an idea neatly encapsulated in John Wheeler's slogan: "It from bit" [matter from information]." *Michael Nielsen* [65]

If one thinks of the mind as an independent "information bearing" entity and the brain as its "processing tool," it is natural to compare the two to the software and hardware of a computer. The role of the "code" in this case would be played by the information that is embedded in the DNA (in compressed form, of course). Such a code would possess the capacity for self-correction in response to interactions with the environment, which is a characteristic that is common to all adaptive systems. It is interesting to note that this analogy actually has some striking similarities with a class of models that are currently used in theoretical biology. Daniel Brooks, for example, has recently proposed a framework for studying genetic systems that closely resembles models used to describe self-correcting algorithms in information theory. [66] Following a similar line of reasoning, biologists John Maynard Smith and Eörs Szatmáry have developed a theory according to which major evolutionary steps tend to coincide with new ways of representing and manipulating information. [67]

It remains to be seen, of course, exactly how effective this approach will prove to be in the study of human consciousness. In making such evaluations, one must bear in mind that consciousness is presumably much more than mere information processing. Indeed, it is fair to say that our feelings, intuition and imagination represent equally important mental functions, which ultimately define us as human beings. It is by no means clear whether these aspects of our consciousness can be adequately described using the concepts of information theory (or physics, for that matter).

In light of the above analysis, I find it difficult to accept the claim that consciousness is reducible to the physical activity of a large number of interconnected neurons. It seems to me that this outlook is overly simplistic, and is bound to miss some important aspects of reality. Many prominent thinkers have shared this view, and have pointed out that materialism cannot possibly capture the human experience in all its uniqueness. Even a very superficial sampling of their opinions conveys a sense of genuine unanimity in this matter.

"Materialism is like a grammar that recognizes only nouns." *Will Durant* [68]

"Not everything that can be counted counts, and not everything that counts can be counted." *Albert Einstein* [69]

"Life is no more made up of physico-chemical elements than a curve is composed of straight lines." *Henri Bergson* [70]

Some Concluding Thoughts

The arguments presented in this section suggest that it is reasonable to believe that the evolutionary process unfolds in the direction of ever increasing complexity. One can also plausibly claim that human consciousness is not strictly material, and that it occupies a special place in the hierarchy of nature. It remains unclear, however, *why* the universe would be organized in such a manner. This obviously isn't the kind of question that science is equipped to answer, but it does fall into the domain of religious discourse. From a theological standpoint, the process of evolution might be viewed as an outpouring of God's infinite love into the world.[71] A loving God who allows a certain degree of freedom to all of creation would not impose His design instantaneously – such an act would be coercive, and would therefore negate the very notion of *agape*. On this view, gradual evolution is necessary for fulfilling the Divine purpose, and increased complexity and consciousness are steps in that direction. It is important to remember, however, that the true goal of this process is presumably a spiritual union with God. The fact that this experience has eluded the great majority of mankind throughout recorded history suggests that creation may be a "work in progress," which requires the constant guidance of a Divine being.

> "The fact is that creation has never stopped. The creative act is one huge continual gesture, drawn out over the totality of time. It is still going on, and incessantly even if imperceptibly, the world is constantly emerging a little further above nothingness." *Pierre Teilhard de Chardin*[72]

> "The long sweep of evolution may not only suggest an unfinished and continuing divine creation but even more radically a creation whose theological status as 'good' may be fully realized only in the eschatological future." *Robert J. Russell*[73]

In support of these views, we might point out that a universe with a definite beginning is much more open to continuous creation than an eternal one. Indeed, something that has existed forever can easily be envisioned as static and unchangeable. In that respect, the Big Bang theory is compatible with the proposed theological explanation, since it implies that the universe has been in existence for a finite amount of time.

The possibility of "creatio continua" is also consistent with recent results in complexity theory, which suggest that nature is inherently open to new forms of organization:

> "Predestiny - or predisposition - must not be confused with predeterminism. It is entirely possible that the properties of matter are such that it does indeed have a propensity to self-organize as far as life, given the right conditions. This is not to say, however, that any particular life form is inevitable. Predeterminism (of the old Newtonian sort) held that everything *in detail* was laid down from time immemorial. Predestiny merely says that nature has a predisposition to progress along the general lines it has. It therefore leaves open the essential unknowability of the future, the possibility for real creativity and endless novelty. In particular it leaves room for human free will." *Paul Davies*[74]

It might be interesting to conclude this section with some thoughts regarding the future development of our species. In recent years, a number of scientists have challenged the role of genes as the primary mechanism of evolution, and have stressed the importance of consciousness in transmitting information from one generation to the next.

> "Human populations no longer adapt to environmental change by evolving genetically. We now adapt by changing and altering our beliefs and behavior." *Anthony Layng* [75]

> "It looks as if the role of natural biological evolution in the foreseeable future will be secondary, for better or for worse, to the role of human culture and *its* evolution." *Murray Gell-Mann* [76]

Such theories are obviously speculative, and have yet to produce convincing empirical evidence. What is interesting, however, is that these scientific conjectures are surprisingly compatible with the views of some religious thinkers. Indeed, some fifty years ago, C. S. Lewis suggested that the next evolutionary form needn't be a 'brainier' human, but rather a more spiritually developed one. [77] This is an intriguing proposition, which is consistent with the thesis that something more than genes may be currently involved in the process of evolution. From that perspective, it would not be irrational to hypothesize that mankind has reached a bifurcation point where the mind is beginning to play a progressively more important developmental role. Genes will still matter, of course, but not exclusively. When it comes to spirituality, evolutionary development ceases to be automatic, and further progress requires the *voluntary* elimination of certain deeply ingrained habits and instincts.

> "The process of spiritual growth is an effortful and difficult one. This is because it is conducted against a natural resistance, against a natural inclination to keep things the way they were, to cling to the old maps and old ways of doing things. ... But in significant numbers humans somehow manage to improve themselves and their cultures. There is a force that somehow pushes us to choose the more difficult path whereby we can transcend the mire and muck into which we are so often born. ... As we evolve as individuals, so do we cause our society to evolve." *M. Scott Peck* [78]

Whether humanity is truly evolving in the direction of spiritual growth is, of course, still an open question. I happen to share some of Peck's optimism in that respect, but this is more a matter of hope than of scientific judgement. Nevertheless, the mere possibility that Darwin's theory may some day be reconciled with the teachings of religious mystics appeals to me both spiritually and aesthetically.

10.4   Is There a Single 'True' Religion?

In one way or another, each religion claims that it possesses a unique knowledge of the truth. Many secular thinkers have argued, however, that this is not logically possible:

> "The wide variety of faiths and their mutual contradictions must mean that at most one of them can be right and that all others are wrong." *Hermann Bondi* [79]

> "It is a matter of logic that, since [different religions] disagree, not more than one of them can be true." *Bertrand Russell* [80]

To Bondi and Russell, the observations articulated above represent sufficient grounds for denying the validity of all forms of theism. Those who believe, on the other hand, question such a conclusion and argue that other interpretations are equally rational. It is important to realize in this context that the debate surrounding competing religious claims is very broad, and includes an entire range of philosophical positions. The extent of this diversity is perhaps best illustrated by the fact that even thinkers who belong to the same religious tradition often have very different perspectives, and find it difficult to reach a consensus.

Among the many proposed responses to the challenge of religious diversity, four deserve particular attention:

1. *Atheism* holds that most religious beliefs are false, and are primarily a result of the human need for a sense of purpose. Claims of religious experiences are typically regarded as projections of our imaginations, and are sometimes even qualified as hallucinations.

2. *Exclusivism* acknowledges the legitimacy of religious belief, but assumes that only one tradition makes correct claims about reality. Most experiences and practices associated with other faiths are dismissed as erroneous and misleading.

3. *Inclusivism* is a more moderate position, which maintains that all religions reflect the Divine reality in some way, but that one particular tradition offers the most direct path to the truth. A typical example of this outlook is the belief that all humans will ultimately find their salvation through Christ, although they may not be directly aware of this. [81]

4. *Pluralism* is a view according to which all traditions essentially represent different "projections" of the same Divine reality. They are assumed to provide equally valid spiritual guidelines and practices, which ultimately lead to a meaningful connection with the "absolute" (note that in some religions the "absolute" is not identified with a Deity).

Over the past few decades, much of the debate concerning religious diversity has focused on the work of philosopher John Hick and his interpretation of pluralism. [82] Hick's main argument revolves around the assumption that there is a single transcendent Real, whose essence is fundamentally unknowable to us. As such, this Real cannot be captured by human concepts, and can only be experienced indirectly, through the different ways in which it interacts with our consciousness. It is important to recognize that the outcome of this interaction is by no means unique, and is conditioned by a range of social, historical and cultural factors.

> "The religious tradition of which we are a part, with its history and ethos and its great exemplars, its scriptures feeding our thoughts and emotions, and perhaps above all its devotional or meditative practices, constitutes a uniquely shaped and coloured "lens" through which we are concretely aware of the Real." *John Hick* [83]

This line of reasoning has lead Hick to conclude that differences in religious teachings, however radical and seemingly irreconcilable, are primarily a product of the *human mind*. While he recognizes that disputed metaphysical issues are clearly important for the way any given religion is practiced, he maintains that having correct beliefs in these matters is *not* a necessary condition for experiencing the Divine reality.

> "The transformation of human existence from self-centeredness to Reality-centeredness seems to be taking place within each of the great traditions despite their very different answers to these debated questions. It follows that a correct opinion concerning them is not required for salvation." *John Hick* [84]

Hick further suggests that the teachings of different traditions are in conflict only when they are perceived as *literal* truths. If they were instead to be understood as purely *mythological* propositions, their truth value could be measured in terms of the response that they evoke in human beings. By that criterion, all religious systems whose myths promote self – transcendence and an appropriate disposition toward the Real would be mutually compatible, and their claims could be viewed as equally valid.

There is a sense in which Hick's outlook bears a resemblance to the theory of relativity. According to Einstein, different observers can record very different times and spatial coordinates in describing the *same* event. For example, what seems to be a stationary object to an observer seated in an airplane will not appear that way to one who is located on the ground. It is fair to say, then, that the *visible* manifestations of physical phenomena depend to a large extent on the chosen frame of reference. And yet, despite the apparent differences, certain *combinations* of the recorded data (such as the so-called *spacetime interval*) remain the same for all observers, as do the basic laws of nature. Special relativity also points to the existence of an "absolute" frame of reference, in which all recorded events become *simultaneous*. From such a vantage point, many of the ambiguities associated with different descriptions of the same reality cease to matter. The only constraint is that this perspective is not available to humans, since material beings cannot travel at the speed of light (see Section 5.1 for more details).

In view of the above discussion, it is tempting to draw an analogy with the world's religions and argue that they are equivalent in certain essential matters (such as basic moral precepts), while differing considerably in the particulars. The apparent discrepancies between these particulars needn't be of primary concern to us, since they do not arise in an "absolute" system of reference (which is presumably the only viewpoint that reflects the true nature of reality). Such an interpretation would clearly be consistent with Hick's version of pluralism, and his contention that genuine spiritual development does not depend on the outcomes of theological disputes.

One can legitimately question, however, whether it is appropriate to dismiss all doctrinal differences as irrelevant (even if we do so only with an "absolute" viewpoint in mind). Indeed, in virtually every major religion there are certain basic teachings whose validity is of paramount importance to all believers. I am inclined to think, for instance, that many Christians would find it extremely difficult (if not impossible) to view the Resurrection as a myth or a simple by-product of the interaction between the unknowable Real and our culturally conditioned minds.

I also feel that a religious tradition should not be valued exclusively for its practical results, however impressive these may be. In order to be meaningful, it must also reflect the *true* nature of reality.

> "Just as I cannot regard science as merely an instrumentally success-ful manner of speaking which serves to get things done, so I cannot regard theology as merely concerned with a collection of stories which motivate an attitude to life. It must have its anchorage in the way things actually are." *John Polkinghorne* [85]

An interesting approach to this problem was proposed by philosopher George Mavrodes, who framed the pluralist dilemma as a choice between two hypothet-ical models. [86] He illustrates the first of these with the story of a prince, who travels through the country in disguise so that he could learn how his subjects live. In one town the prince is dressed as a monk and in another as a stonema-son, thus presenting very different appearances to the people who see him. But there is nothing fictitious about either of these experiences - both are genuine, and refer to a real source (the monk, the stonemason and the prince are, in fact, the *same* person).

In the second model, Mavrodes asks us to imagine several artists painting a landscape. Although their work is entirely abstract and bears little resemblance to the actual objects that are depicted, it is fair to say that each painting is influenced by the landscape, and represents a legitimate reflection of reality. We should add, however, that there is also a decidedly human contribution to the images, which derives from the aesthetic sensibilities of the individual artists. It would therefore be incorrect to hold that the paintings and the landscape are the *same* thing.

The question that Mavrodes raises with the help of these parables is whether we should think of different religious beliefs as valid appearances of the same reality (as in the case of the disguised prince), or whether we ought to interpret them as images produced by the human psyche in response to some kind of intuitive awareness of the Real (as suggested by the metaphor with the artists). The latter possibility has potentially damaging consequences for established religions, since it implies that some of their central teachings may be no more than myths, whose relationship to the Real is indirect at best. It is therefore very important to assess which of these two positions has more credibility.

In dealing with this difficult issue, an analogy with quantum mechanics may help bring some clarity. To properly understand the connection, we should first recall that light can behave *both* like a wave *and* like a collection of particles, depending on the kind of measurement that it is exposed to (this is known as the Principle of Complementarity). Although the two features appear to be mutually exclusive in our mundane experience, modern physics nevertheless permits them to coexist, and treats them as ontologically equivalent (it makes no sense to say that light is "more" a wave than a particle). A similar situation arises in the case of quantum particles, whose "state" prior to a measurement consists of many alternatives that are never encountered together in the physical world that we inhabit. Of course, once the measurement is performed, only one of these alternatives is detected (by virtue of the "collapse" of the wave function).

To this we might add the observation that science often allows for the coexistence of multiple explanations, even in cases when they appear to contradict each other. A typical example of such a situation are the theories proposed by Max Born and Richard Feynman, which pertain to the nature of the "waves" that are associated with quantum particles (see Section 4.3 for more details). Although these two models represent radically different interpretations of reality, they are known to produce equally accurate predictions. Since the essence of the phenomenon they are describing is unknown to us, we have no grounds to say that one is "true" and the other is not.

If we were now to draw an analogy with religious diversity, it seems to me that the comparison would favor a version of the "disguised prince" model in which *each* of the different appearances represents an accurate manifestation of some aspect of reality. Indeed, the fact that seemingly incompatible theories are allowed to coexist in science suggests that a single phenomenon can be described in multiple ways, and that all these descriptions are legitimate despite being contradictory by human standards. In explaining such occurrences, there is no need to appeal to cultural conditioning or the workings of our collective imaginations – this is simply a consequence of the fact that truth is complex and multidimensional.

The above discussion suggests that it is not irrational to believe that different traditions can provide genuine experiences of the Divine reality, despite the apparent discrepancies in their teachings. But if we accept this hypothesis, we must also consider whether all such accounts should be viewed as *equally* meaningful. It is important, in other words, to objectively evaluate pluralist claims, and examine whether it is justified to hold that the teachings of one religion might be more accurate than those of others. If this turns out to be the case, it could be reasonably argued that "soft" inclusivism is a defensible position.

We begin by observing that both quantum mechanics and the theory of relativity are limited to finite entities, and descriptions of their behavior in the framework of space and time. Given that religious claims often transcend finite and empirically observable phenomena, a more abstract analogy from mathematics might bring some additional insights. In order to formulate such an analogy, let us recall that any vector in a three dimensional space can be represented as

$$v = \alpha_1 e_1 + \alpha_2 e_2 + \alpha_3 e_3 \qquad (10.2)$$

where $\{e_1, e_2, e_3\}$ denotes unit vectors along the three axes (the so-called basis of the space), and $\{\alpha_1, \alpha_2, \alpha_3\}$ are the projections of v. This notion can be extended to infinite dimensional spaces, whose elements can be expressed in the form

$$v = \sum_{i=1}^{\infty} \alpha_i e_i \qquad (10.3)$$

When the basis e_i ($i = 1, 2, \ldots$) is fixed for a given space, the set of projections $\{\alpha_1, \alpha_2, \ldots\}$ fully describes vector v.

Without getting into further details, it suffices to observe that we must confine any practical description of vector v to a *finite* subset of its projections. The challenge, then, is to determine which subset constitutes the 'best' approximation of the infinite dimensional object. In a certain sense, it could be said that each finite subset of projections is 'correct', and describes reality in its own way. This does not mean, however, that all approximations are equally accurate. In the case of Fourier series, for example, a small (and very well defined) subset of projections specifies the original function well, while other choices do not.

The key question in this context is how one can identify a good approximation when the infinite dimensional entity is *unknown*. To see how that might be done, let us suppose that our existing knowledge about v suggests that this vector can be reasonably approximated with a finite number of terms, as [87]

$$\hat{v} = \sum_{i=1}^{n} \hat{\alpha}_i e_i \qquad (10.4)$$

Following the standard approach of functional analysis, we will introduce the "distance" between v and $\hat{v}$ as

$$d = \|v - \hat{v}\| \qquad (10.5)$$

where $\|\cdot\|$ represents an appropriate norm in the infinite dimensional space. [88] In principle, any new information that we acquire about v will lead to a better approximation

$$\tilde{v} = \sum_{i=1}^{m} \beta_i e_i \qquad (10.6)$$

which consists of m terms (with $m > n$). However, if the original distance $d = \|v - \hat{v}\|$ happens to be sufficiently small, the additional $m-n$ terms in (10.6) will have a minimal impact on the overall sum. In other words, the emergence of new information about v will *not* require significant changes in the original estimate (10.4). We can detect such situations without knowing v explicitly, by simply monitoring the difference $\|\tilde{v} - \hat{v}\|$. When this difference drops below a certain preassigned threshold value, it is reasonable to say that the process of approximating v has *converged*.

If we were now to imagine God as an infinite dimensional being and different religious descriptions of His essence as finite 'subsets of projections', an analogy with our mathematical example would suggest three conclusions:

1. All religious traditions can be viewed as valid (although necessarily incomplete) approximations of the Divine reality.

2. The descriptions of God provided by different belief systems needn't be equally accurate reflections of the truth.

3. Different belief systems might be compared by evaluating how they respond to new knowledge about observable reality. If, for example, the teachings of a particular tradition require little or no modification as new scientific information becomes available, this could be viewed as an indicator of some sort of "convergence" (at least when it comes to the physical manifestations of reality). It would then be reasonable to claim that such a system reflects the truth more accurately than traditions whose basic postulates contradict verifiable facts.

In light of the conclusions outlined above, it would appear that there is no single 'true' religion, but that some traditions possibly embody the truth more directly than others. It might even be argued that our theoretical and practical understanding of physical reality can serve as a possible tool for establishing this distinction. Such an outlook is obviously quite different from Hick's, and seems to allow some room for religious inclusivism. We should be careful, however, not to take this analogy too far. In mathematics diversity needn't be a good thing, particularly if we know that one approach is more accurate than others. In matters of faith, on the other hand, the situation is quite different:

> "Is the existence of so many religious types and sects and creeds regrettable? I answer no emphatically ... Each attitude being a syllable in human nature's total message, it takes the whole of us to spell the meaning out completely." *William James* [89]

Along these lines, Huston Smith points out the ultimate consistency of all spiritual traditions:

> "In searching for God, people will normally follow the path dictated by their own civilization. This is fine. However, those who circle the mountain trying to bring others to their paths are not climbing. It is, ultimately, possible to climb from all sides, although some paths

are likely to be more direct than others. At the top, however, all trails converge." *Huston Smith*[90]

The above discussion suggests that it is perfectly sensible to view the diversity of religious traditions in a positive light, even if we choose to adopt an inclusivist position. The many existing differences needn't be interpreted as a sign of epistemic unreliability, and can instead be seen as a natural consequence of human freedom (which a loving God necessarily allows to all of creation). However, even if this is the case, certain important issues related to the rationality of faith remain unresolved. Namely, it has often been argued that our religious affiliation is determined by the circumstances of our birth, which we cannot control. But if we agree that all traditions reflect the truth in some way, why would we have to believe the claims of a particular religion that was 'imposed' on us by chance? Shouldn't we rather choose our faith *freely*, based on rational criteria?

In thinking about this question, it might be useful to make a comparison with the way we learn to speak. Given the environment into which we are born, our native language happens to be the most appropriate form of communication. It is true that this language is imposed on us, but we must also acknowledge that it fits best with the social environment and the concepts required to live in a particular time and place. Eskimos of Western Greenland, for example, have some 49 words to describe different kinds of snow and ice, but their language would hardly be suitable in the tropics. Woody Allen might have had something like this in mind when he once jokingly remarked that in the language of his grandmother there was no word for love, but 15 words for headache.

Like language, religion has certain rules and a logic that is accepted at an early age. It is important to recognize, however, that these rules are not simply conventions that are to be accepted blindly. They are, in fact, a result of optimized human experience, collected and formed in a particular environment over thousands of years. As such, basic religious beliefs and traditions are naturally adapted to given cultural and historical settings.

Although our native beliefs are deeply ingrained in the way we think, they do not automatically prevent us from exploring other forms of spirituality. This can be an enriching process, in which we broaden our horizons and at the same time gain a deeper understanding of our own tradition. But in the end, the ideas and concepts that we grew up with generally tend to remain our natural mode of understanding – this is as true of religious beliefs as it is of languages. There is nothing irrational in adhering to either, despite the fact that we never freely chose them.

One might even question whether such choices can ever be completely free and rational. When notions such as love and faith enter the picture, the very nature of the decision making process changes. As an illustration, suppose that you are happily married, and that the relationship involves a degree of genuine selfless love. Suppose further that you are asked how you can be sure that somewhere in the world there isn't another person that perhaps better matches your temperament and interests. It seems to me that from a purely logical standpoint you cannot claim that such a person does not exist. However, a loving relationship entails far more than analytical thinking. In response to such a question one could argue that who we are *now* is so conditioned by the person we love that there really is no better match to be found. Such a match may have been possible at the outset of the relationship, but not after it has evolved to a sufficient degree. In other words, once you have transcended your "self" and given up part of it to another, the choice becomes essentially unique.

We might add that in the case of religion, the possibility of a purely rational choice comes rather late in life. Most people are introduced to a particular tradition at birth, and by the time they begin to ponder it objectively, this tradition is already integrated into their personality. As in the case of love, the more genuine and selfless the faith is, the lower the likelihood that there exists an adequate alternative.

10.5 Notes

1. Kitty Ferguson, *The Fire in the Equations*, W. B. Eerdmans Publishing Co., 1994.

2. Huston Smith, *The World's Religions*, Harper San Francisco, 1991.

3. C. S. Lewis, *Miracles*, Harper Collins, 2001.

4. George Smith, *Atheism: The Case Against God*, Prometheus Books, 1989.

5. www.worldofquotes.com

6. John Barrow, *Impossibility: The Limits of Science and the Science of Limits*, Oxford University Press, 1998.

7. Bertrand Russell, *Religion and Science*, Oxford University Press, 1997.

8. David Hume, *An Enquiry Concerning Human Understanding*, Pearson Education, 1955.

9. Thomas Paine, "The Age of Reason," in *Thomas Paine: Collected Writings*, The Library of America, 1995.

10. See, *e.g.*: Michael Mahoney, *Scientist as Subject*, Ballinger, 1976.

11. Ibid.

12. Unfortunately, I cannot recall where I read this.

13. The word 'almost' is permissible even in mathematics. For example, we say that a function is differentiable almost everywhere if it has this property in the entire domain with the exception of certain isolated points. It is interesting to note in this context that the probability of running into such a point is zero, but it is certainly possible!

14. Paul Davies, "Teleology Without Teleology: Purpose through Emergent Complexity," in *Evolutionary and Molecular Biology*, R. J. Russell, W. R. Stoeger and F. J. Ayala (Eds.), Vatican Observatory Publications, 1998.

15. Ferguson, *The Fire in the Equations*.

16. Paul Kurtz, "Examining Claims of the Paranatural," in *Science and Religion: Are They Compatible?*, Paul Kurtz (Ed.), Prometheus Books, 2003.

17. Quoted in: Ferguson, *The Fire in the Equations*.

18. Norwood R. Hanson, *Perception and Discovery*, Freeman Cooper, 1969.

19. There are numerous examples of this kind in the history of science, including the very recent past. One of the most commonly cited ones concerns the so – called "neutral currents," which are associated with the weak nuclear force. Although experimental evidence for this phenomenon was available in the early 1960s, it was repeatedly dismissed until Abdus Salam and Steven Weinberg proposed a theory that unified the electromagnetic and weak forces.

20. http://en.wikiquote.org, 2005.

21. Richard Purtill, *Thinking About Religion: A Philosophical Introduction to Religion*, Prentice-Hall, 1978.

22. Quoted in: Ian Barbour, *When Science Meets Religion*, Harper San Francisco, 2000.

23. Peter Godfrey-Smith, *Theory and Reality*, University of Chicago Press, 2003.

24. http://en.wikiquote.org, 2005.

25. Ibid.

26. Ernan McMullin, "Long Ago and Far Away: Cosmology as Extrapolation," in R. Fuller (Ed.), *Bang: The Evolving Cosmos*, University Press of America, 1994.

27. Quoted in: S. Chandrasekhar, *Truth and Beauty: Aesthetics and Motivations in Science*, University of Chicago Press, 1987.

28. Isaiah Berlin, "The Concept of Scientific History," in *The Proper Study of Mankind*, Farrar, Straus and Giroux, 2000.

29. A good discussion of the nature scientific explanations can be found in: Thomas Kuhn, *The Essential Tension: Selected Studies in Scientific Tradition and Change*, University of Chicago Press, 1977.

30. Aldous Huxley, *The Perennial Philosophy*, Harper and Row, New York, 1972.

31. Ibid.

32. Quoted in: Martin D'Arcy, *St. Thomas Aquinas*, Newman Press, 1955.

33. Quoted in: Diane Ackerman, *An Alchemy of Mind: The Marvel and Mystery of the Brain*, Simon and Schuster, 2004.

34. Quoted in: Huxley, *The Perennial Philosophy*.

35. Quoted in: Dorothee Soelle, *The Silent Cry: Mysticism and Resistance*, Fortress Press, 2001.

36. Smith, *Atheism: The Case Against God*.

37. Bertrand Russell, *The ABC of Relativity*, Mentor, 1985.

38. Quoted in Leonard Shlain, *Art and Physics*, Harper Collins, 1991.

39. Philip Kitcher, "Explanatory Unification and the Causal Structure of the World," in *Scientific Explanation*, P. Kitcher and W. Salmon (Eds.), University of Minnesota Press, 1989.

40. David Bartholomew, *Uncertain Belief*, Oxford University Press, 2000.

41. http://en.wikiquote.org, 2005.

42. Smith, *The World's Religions*.

43. Ibid.

44. William Spohn, *Go and Do Likewise*, Continuum, 1999.

45. Bertrand Russell, *A History of Western Philosophy*, Simon and Schuster, 1972.

46. St. Thomas Aquinas, *Summa Theologiae*, Part I, Q. 2.

47. John Paul II, *Truth Cannot Contradict Truth*, in a statement to the Pontifical Academy of Sciences, October 22, 1996.

48. I tend to think of a theological paradigm as a method for deriving specific teachings from the articles of faith. In that respect, a "paradigm" plays the same role that "rules of inference" play in a formal axiomatic system. When using this analogy, it is important to remember that changes in the rules of inference impact *only* the theorems, while the axioms remain unaffected. With that in mind, it is entirely reasonable to hold that the *same* basic principles can be compatible with an unlimited number of religious paradigms.

49. Max Planck, *Scientific Autobiography and Other Papers*, Greenwood Publishing Group, 1968.

50. Thomas Kuhn, *The Structure of Scientific Revolutions*, University of Chicago Press, 1996.

51. Richard Dawkins, *The Blind Watchmaker*, W.W Norton & Co., 1986.

52. Stephen Jay Gould, *Ever Since Darwin*, W.W. Norton & Co., 1977.

53. Paul Davies, *The Cosmic Blueprint*, Simon and Schuster, 1988.

54. William Stoeger, "The Immanent Directionality of the Evolutionary Process, and its Relationship to Teleology," in *Evolutionary and Molecular Biology*, R. J. Russell *et al.* (Eds.).

55. Quoted in: Ethan Watters, "DNA Is Not Destiny," *Discover Magazine*, November 2006.

56. Robert Russell, "Special Providence and Genetic Mutation: A New Defense of Theistic Evolution," in *Evolutionary and Molecular Biology*, R. J. Russell *et al.* (Eds.).

57. Charles Darwin, *The Origin of Species*, Oxford University Press, 1998.

58. Quoted in: Peter van Inwagen, "Quam Dilecta," in *God and the Philosophers*, Thomas V. Morris (Ed.), Oxford University Press, 1994.

59. Autocatalysis is a nonlinear process in which chemical substances accelerate the reactions by which they are formed. For more details, see *e.g.*: Prigogine and Stengers, *Order Out of Chaos*, Bantam Books, 1984.

60. See, *e.g.*: Per Bak, *How Nature Works: The Science of Self Organized Criticality*, Springer-Verlag, 1996. A more rigorous mathematical treatment of this topic can be found in: Henrik Jensen, *Self-Organized Criticality*, Cambridge University Press, 1998.

61. N. Eldredge and S. J. Gould, "Punctuated Equilibria: An Alternative to Phyletic Gradualism," in *Models in Paleobiology*, T. J. Schopf (Ed.), Freeman, 1972.

62. Francis Crick , *The Astonishing Hypothesis: The Scientific Search for the Soul*, Charles Scribner & Sons, 1994. Crick was one of the discoverers of the DNA double helix structure.

63. Bartholomew, *Uncertain Belief*.

64. Quoted in: John Haught, *Is Nature Enough?*, Cambridge University Press, 2006.

65. Quoted in: Gregory Chaitin, *Meta-Math: The Quest for Omega*, Pantheon Books, 2005. A quantum mechanical interpretation of this outlook can be found in: Evan Walker, *The Physics of Consciousness*, Basic Books, 2000.

66. Daniel Brooks "Evolution in the information age: Rediscovering the nature of the organism," *Semiosis, Evolution, Energy, Development*, 1, 2001.

67. John Maynard Smith and Eörs Szatmáry, *Major Transitions in Evolution*, Oxford University Press, 1998.

68. Will Durant, *The Story of Philosophy*, Simon and Schuster, 1972.

69. www.worldofquotes.com

70. Henri Bergson, *Creative Evolution*, Dover, 1998.

71. John Haught, *Science and Religion: From Conflict to Conversation*, Paulist Press, 1995.

72. Pierre Teilhard de Chardin, *The Prayer of the Universe*, Harper & Row, 1968. Teilhard de Chardin was a famous Jesuit paleontologist.

73. Russell, "Special Providence and Genetic Mutation: A New Defense of Theistic Evolution."

74. Davies, *The Cosmic Blueprint*.

75. Anthony Layng, "Supernatural Power and Cultural Evolution," in *Science and Religion: Are They Compatible?*

76. Murray Gell-Mann, *Quark and the Jaguar: Adventures in the Simple and the Complex*, Henry Holt & Co., 1995.

77. C. S. Lewis, *Mere Christianity*.

78. M. Scott Peck, *A Road Less Traveled*, Simon and Schuster, 1998.

79. Hermann Bondi, "Uniting the World – Or Dividing It?" in *Science and Religion: Are They Compatible?*

80. Bertrand Russell, *Why I Am Not a Christian and Other Essays on Religion and Related Subjects*, Simon and Schuster, 1976.

81. The famous theologian Karl Rahner referred to such individuals as "anonymous Christians."

82. A comprehensive exposition of Hick's views on religious diversity can be found in: John Hick, *An Interpretation of Religion*, Yale University Press, 2004.

83. John Hick, "Religious Pluralism and Salvation," in *The Philosophical Challenge of Religious Diversity*, Philip Quinn and Kevin Meeker (Eds.), Oxford University Press, 2000.

84. Ibid.

85. Polkinghorne, *The Faith of a Physicist*.

86. George Mavrodes, "Polytheism," in *The Rationality of Belief and the Plurality of Faith*, Thomas Senor (Ed.), Cornell University Press, 1995.

87. This type of situation occurs, for example, in interpolation problems, where we know only n points of a function $f(x)$. In that case, the best we can do is approximate $f(x)$ as

$$\hat{f}(x) = \sum_{k=0}^{n-1} \hat{\alpha}_k x^k \tag{10.7}$$

where the polynomials $\{1, x, x^2, \ldots, x^{n-1}\}$ correspond to the basis previously denoted by $\{e_1, e_2, e_3, \ldots, e_n\}$.

88. A norm can be viewed as a measure of distance between two objects in a linear space. The distance introduced in (10.5) is defined but cannot be computed, since v is assumed to be unknown.

89. William James, *The Varieties of Religious Experience*, Dover Publications, 2002.

90. Smith, *The World's Religions*.

Chapter 11

Epilogue

In his book *Science and Religion: From Conflict to Conversation*, theologian John Haught outlined four distinct ways in which science and religion can relate to each other. The different possibilities (which he identified as *conflict*, *contrast*, *contact* and *confirmation*) can be described as follows:

1) The *conflict* approach suggests that science and religion have irreconcilable differences, and that there is no room for compromise. From that standpoint, only one of these two disciplines can be a faithful reflection of the truth.

2) The *contrast* approach claims that science and religion address very different questions, and that there is no meaningful overlap between them. That being the case, both disciplines ought to refrain from making pronouncements about matters that lie outside of their domain of inquiry.

3) The *contact* approach holds that there is enough common ground for a meaningful dialogue, since both science and theology provide an incomplete understanding of reality. On this view, the two disciplines should be aware of each others' methods and achievements, and should use this knowledge whenever possible.

4) The *confirmation* approach claims that science and religion reinforce each other in subtle ways. Since both disciplines reflect the same ultimate reality, there is a genuine possibility that the two might be able to cooperate on some level.

In light of this classification, it seems appropriate to close our discussion by considering which of the four possibilities carries the most weight. We might begin with a brief critique of the "conflict" position, since it is clearly the least constructive approach to the debate. Those who adopt such an outlook tend to base their arguments on perceived certainties, and allow little (if any) room for doubt regarding the validity of their claims. Although this has been (and still is) a widespread attitude among believers and skeptics alike, I would argue that it has some serious methodological deficiencies. Indeed, if we acknowledge that the scientific method (and even logic itself) have certain inherent limitations, and that our understanding of God is necessarily incomplete, we *must* allow for the possibility that all claims about the ultimate nature of reality contain an element of irreducible uncertainty. Refusal to recognize this possibility cannot be justified on rational grounds, and can only be attributed to certain deeply ingrained psychological tendencies.

Given the potential dangers and shortcomings associated with unshakable beliefs, should we then all become agnostics when it comes to religion and its teachings? This seems like an entirely reasonable prospect, particularly if we assume that agnosticism spans the entire spectrum of beliefs that lie between biblical literalism and radical atheism. If we adopt such a position, the key question becomes whether all the possibilities in this range have *equal* logical and experiential justification.

It goes without saying, of course, that evaluations of this sort are notoriously difficult to make, since there are no obvious objective criteria for comparing the different alternatives (apart from logical consistency). An intuitively appealing option would be to adopt a position that is somewhere near the middle of the "agnostic spectrum," perhaps with a slight preference in one direction or the other. Such a "neutral" philosophical outlook appears to be intellectually safe, and is a reasonable reflection of the level of uncertainty that surrounds fundamental theological claims. We should exercise caution, however, when it comes to extending this type of reasoning to the domain of religion. Indeed, religion in general (and Christianity in particular) demand a firm commitment to a set of basic beliefs and a certain way of life, *despite* the apparent lack of conclusive evidence. Christian theology acknowledges that most believers will probably have their share of doubts, but maintains that these are outweighed by the potential benefits.

> "If belief in God and Christian doctrine were a purely intellectual matter of no practical consequence, then a more or less permanent state of agnosticism would be appropriate. ... But this is not the case here. It would be a shame to miss the whole point of living because it took a lifetime, and more, to decide what it was." *David Bartholomew* [1]

If we agree, then, that faith requires us to abandon neutrality in matters of theological importance, where does this leave scientifically minded individuals who are open to religion? Much of the argument in the preceding chapters was aimed at showing that such people *can* make the necessary transition without compromising their standards of rationality. What remains unclear, however, is which position they ought to adopt in the debate between science and religion. Should they remain cautious and treat these two domains as completely separate (as the "contrast" approach suggests)? Or should they embrace the more optimistic view that conversation and perhaps even cooperation are possible? In my opinion, the answer to this question ultimately becomes a matter of personal inclination. As long as one is consistent and logical, each of the last three options on Haught's list can be rationally justified. I therefore offer no clear cut answer, and can only share my own views on the subject, which tend to favor the "confirmation" approach.

I cannot say in all honesty how and when I adopted this position - I can only note that it was a lengthy process, which involved a great deal of introspection. The biggest obstacle that I had to overcome was probably my "initial condition," which was not particularly favorable toward religion. In my college days I was actually quite skeptical toward theological claims, and was inclined to give science a decided advantage in all matters in which the two disciplines conflicted. It is fair to say, however, that my opinions have changed significantly over the past twenty five years. I am tempted to attribute this change to the fact that I am now older (and possibly wiser), but I am not sure that this would be the best explanation. Looking back, it seems to me that the shift in my attitude toward religion probably had more to do with my long-standing fascination with beauty and harmony, both in nature and in art. I have always felt that there is something truly mysterious and profound in the fact that aesthetic considerations can (and often do) lead to deep and unexpected truths. This didn't seem like something that science could explain, so I began looking elsewhere.

As I pondered this question, I considered, of course, the possibility that our "physical" experience of beauty may be no more than a complex set of neural interactions in the brain. If this were to be conclusively established some day, would such a finding remove the "sense of beauty" from its pedestal in the hierarchy of human values, and reduce it to just another physiological process? I seriously doubt that. It seems highly unlikely that any scientific model, whether analytical or empirical, will be able to explain how the human imagination can reach beyond the "merely observable," and intuitively anticipate truths about the nature of reality. Indeed, while it is reasonable to expect that biochemical processes can have survival value and can even lead to various forms of self-organization, we have no grounds whatsoever to expect them to produce the likes of quantum mechanics or the paintings of Vermeer. This mysterious capacity of the human mind to create and provide meaningful insights suggests to me that an even greater creative mystery lies behind it, a *single* transcendent reality of which we are perhaps only pale reflections. It would be reasonable to assume that this reality manifests itself in different ways, and allows for multiple modes of inquiry (including, of course, both science and theology).

The claim that science and theology relate to different aspects of the same "ultimate truth" finds additional support in the fact that both disciplines are built around the premise that the universe is organized in a lawful manner, and that it is open (at least in part) to our methods of investigation. Indeed, without an a priori belief that some *discernible* organizing principles exist, we probably never would have bothered to look for them in the first place. It is highly unlikely that science (at least, as we know it today) could have developed under such circumstances. This is not to say, of course, that humans would have failed to observe and utilize certain patterns in nature without a set of prior beliefs. They most probably would, since pattern recognition has an obvious survival value. We should point out, however, that there is a big difference between observing regularities in nature, and the expectation that there is a unifying explanation for them. This expectation that the universe is somehow "intelligible" to us is ultimately an axiomatic proposition, which found its expression in religion long before the emergence of modern science. From that perspective, it would not be inappropriate to say that the scientific enterprise as a whole is motivated by certain basic philosophical principles which are shared by theology.

In further evaluating the possibilities for "cooperation" between science and religion, it is useful to reiterate that both disciplines recognize their limitations, and can even anticipate what kinds of questions they will never be able to answer. It is in this domain that science and theology might find some common ground, and room for genuine interaction. Along these lines, philosopher Ernan McMullin pointed out that the limits of science provide an opportunity for genuine self-transcendence and intuitive "leaps" from the known into the "realm of the invisible". This type of inference (which is traditionally associated with theology) is perhaps most apparent in the domain of cosmology, where we routinely make extrapolations from our own limited experience to the universe as a whole, despite the fact that it eludes our methods of observation.

> "Modern cosmology has been ... directed by 'principles' whose credentials are remarkably difficult to assess. The degree of conceptual extrapolation is so extreme and the possibilities of empirical testing are so slender that cosmologists often have to rely on the most elusive of intuitions. ... These intuitions derive from sources that in many cases lie outside the confines of 'normal' science. This is why the boundaries between cosmology and metaphysics or even theology seem so permeable." *Ernan McMullin* [2]

In drawing such parallels between science and theology, we would be well advised to avoid the so-called "God-of-the-gaps" arguments, which justify the existence of God by identifying phenomena that have not yet been explained. Claims of this sort are not taken very seriously in philosophical literature, since they can be (and often are) falsified by new scientific discoveries. If one wishes to pursue this line of reasoning, it would be far more productive to focus on the kind of fundamental unknowability that follows from science itself. Results such as Gödel's Incompleteness Theorem and Heisenberg's Uncertainty Principle, for example, clearly indicate that certain aspects of reality will always remain beyond the reach of science (a view that is shared by theology as well). Some of our most profound experiences as human beings support this conclusion, and point to the fact that there is a great deal of knowledge to be found beyond the domain of mathematical models and empirical testing. This knowledge may be extraordinarily complex and subtle, but should not be viewed as inaccessible. Hints regarding its nature can actually be found everywhere around us - in the love we feel for others, in human compassion and solidarity, in art and beauty, and in the fact that human imagination can sometimes penetrate the deepest and most hidden secrets of the physical universe without even a shred of experimental evidence. All this seems to be consistent with the existence of a greater organizing principle, which we can reasonably associate with a benevolent God if we feel the inclination to do so.

If we allow, then, for the possibility that science and religion are reflections of the same unified reality, how exactly might these two approaches to the "truth" collaborate *in practice*? A partial answer to this question coincides with one of the principal objectives of this book, which has been to show that science can provide "analogical bridges" between the two disciplines. This, however, seems to be a somewhat lopsided argument, which assigns a passive (and almost subsidiary) role to theology. In order to restore a measure of "symmetry" in this interaction, we would presumably have to show that theology can be of some use to scientists in their everyday work (which does not seem to be a very likely prospect). I was actually asked this very question by a thoroughly secular friend of mine, to whom I am indebted for raising the issue with such precision. My response was that faith has little bearing on *how* I go about my research, but that it most certainly affects *why* I do it in the first place. It would be wrong, in other words, to look for the practical impact of theology in the process of scientific investigation per se - its true value lies in the added meaning and motivation that it can provide to the overall enterprise.

I cannot say that I arrived at this conclusion by way of a sophisticated philosophical argument - it was mainly the result of my personal experiences, both in science and beyond. For the most part, these experiences were neither momentous nor particularly revelatory. And yet, they had a cumulative effect on my attitude toward research that was far from trivial. I clearly recall one such event, which occurred around the time when I was applying for tenure. I was driving back from work one evening when I noticed that the car in front of me had a bumper sticker which read: "She who dies with the most shoes wins." My immediate reaction, of course, was to laugh at the joke, but there was also a delayed reaction, which became apparent only a day or two later. I began to wonder, first subconsciously and then consciously, whether my drive to publish as many papers as possible was all that different from what the bumper sticker was ridiculing. There is no doubt, of course, that a large number of publications can be extremely helpful (perhaps even decisive) when it comes to furthering one's academic career. But this paradigm clearly becomes counterproductive when professional recognition becomes the primary motive for research, and the thrill of discovery is replaced by sheer expediency. Although I had not reached such a point at the time, this incident brought the problem to my attention with unexpected clarity.

Over time, this undoubtedly trivial event became a sort of "symbolic reference point" for other decisions, which ultimately led to a significant shift in my attitude. From that perspective, it would be fair to say that this incident had a more direct impact on me than any philosophical argument that I read on the subject. This does not mean, of course, that the consequences were immediate - it actually took years before I noticed any concrete changes in my approach to research. As is the case with most important transformations in life, this one began quietly and almost accidentally, and slowly worked its way from the heart to the mind. I don't think there is anything particularly unusual about such experiences - at one time or another, all of us go through a similar process. This is how we grow to love others, and ultimately learn to transcend our self-interest. And this is also how faith and spirituality make their way into our lives. As St.

Augustin said many centuries ago, it is not theoretical explanations, but rather personal experience that is the proper path to God.

> "Need it concern me if some people cannot understand this [relationship between man and God]? Let them ask what it means, and be glad to ask: but they may content themselves with the question alone. For it is better for them to find You and leave the question unanswered than to find the answer without finding You." *St. Augustin* [3]

Notes

1. David Bartholomew, *Uncertain Belief*, Oxford University Press, 2000.

2. Quoted in: Paul Allen, *Ernan McMullin and Critical Realism in the Science-Theology Dialogue*, Ashgate, 2006.

3. St. Augustin, *Confessions*, Barnes and Noble Classics, 2007.

Index

agape, 132, 166, 168, 214
analogies
    and biology, 207
    and ethics, 128–130, 132
    and mathematics, 7
    and plausibility of beliefs, 7, 9
    and quantum mechanics, 51
    and theology, 6, 155, 164, 167,
        175, 202, 217, 219
Aquinas, St. Thomas, 12, 130, 154,
        161, 163, 172, 177, 198, 204
Augustine, St., 130, 166, 177
axioms
    as elements of formal systems, 36
    in geometry, 34, 35
    in theology, 155, 156

beauty
    and mathematical order, 106,
        108, 116
    and minimalism, 112, 113
    and simplicity, 5, 103
    and symmetry, 96, 102
    and truth, 101, 104
    as unity in diversity, 114
    evolutionary origins of, 95
    theological interpretations of,
        96–98, 100, 105
Bell's inequality, 63
Big Bang
    and the spacetime singularity, 72,
        149
    background radiation from, 73
    theological implications of, 192,
        214
black hole
    and unknowable truths, 151
    event horizon of, 76

    evidence for, 77–79
Bohr, Niels, vi, 51, 53, 58, 60, 64, 159,
    162, 194
Boolean networks, 28, 29, 209
butterfly effect, 21

Cantor, Georg, ix, 112, 164
chaos
    and intermittency, 19, 190
    and sensitivity to initial
        conditions, 18, 21, 22
    and strange attractors, 22, 23
    and unpredictability, 18, 22, 146
coloring problem, 114
complexity
    and emergent phenomena, 100,
        167, 211
    and theology, 175
    as a condition for information
        processing, 30
    in algorithmic information
        theory, 26, 152
    in dynamic systems, 23, 27

Darwin, Charles, 185, 201, 207, 210,
    212
determinism
    and chaos, 15
    and free will, 127
    and predictability, 16, 18, 19
    and quantum mechanics, 54

Einstein, Albert, vi, 5, 52, 60, 62, 65,
    67, 68, 70, 72, 95, 103, 104,
    114, 130, 147, 161, 177, 187,
    193, 213
EPR paradox, 60, 62
ethics

and free will, 126
and moral relativism, 130, 134
and psychology, 139, 140
evolutionary origins of, 126
in scientific practice, 135–137
religious interpretations of, 130,
    132
Euclid, 34, 113, 155
evil
    and divine omnipotence, 174
    and free will, 169, 174
    and freedom of process, 170
evolution
    and complexity, 206
    and consciousness, 211, 213
    and information, 209, 213
    and self-organization, 208, 211
    and theology, 207, 210, 214
    as a random process, 207, 209

Fermat's Last Theorem, 41
formal systems, 35
fractals
    and Pollock's paintings, 24
    and strange attractors, 23
    and the Koch curve, 25
    dimensionality of, 23, 24

general relativity
    and curved spacetime, 70
    and Newton's theory of gravity,
        69
    experimental evidence for, 71
God
    and infinity, 164, 165, 180, 221
    as omnipotent, 172
    as omniscient, 176
    as personal, 164
    as self-limiting, 174, 179
    as suffering, 171
    as timeless, 177, 179
    as transcendent, 145
    as unconditional love, 166
    descriptions of, 10, 160, 161, 165
    existence of, 154, 157
Godel, Kurt, 33, 38–40, 44, 151, 156,
    158, 173

golden ratio, 117, 118, 120

Heisenberg, Werner, 51, 54, 56, 57, 60,
    83, 104, 105, 147, 171, 194
Hick, John, 155, 216, 221

Incompleteness Theorem
    and unprovable propositions, 38
    implications of, 40, 151, 232
interactions
    and complexity, 30
    and theology, 167
    between autonomous agents, 27,
        28
internal consistency
    in Euclidean geometry, 34
    in formal systems, 37
    of religious beliefs, 6

Jesus Christ, 171, 191, 203, 204, 216
Jung, Carl, 99, 168

Kauffman, Stuart, 28, 209
Kuhn, Thomas, 136, 205

Leibniz, Gottfried Wilhelm von, 104,
    125, 139, 176

Mandelbrot set, 45
metamathematics, 33
miracles
    and laws of nature, 190
    as a result of ignorance, 187
    as improbable, 188
    evidence for, 191

Newton, Isaac, 58, 68, 69, 75, 102,
    135, 139, 190

Omega, 152

particle
    families, 84
    spin of, 60
    virtual, 84
particle-wave duality, 58

Polkinghorne, John, 103, 167, 170,
      172, 176, 178, 218
Principia Mathematica, 38, 39
probability
      and dispersion, 55, 56
      Bayes' theorem, 8
      distribution, 55, 56

quantum mechanics
      and general relativity, 83, 148
      and the many worlds
            interpretation, 54, 159
      and the measurement problem,
            53
      and the state of superposition,
            53, 60, 61

rationality
      of belief system, 10, 157
      standards of, 4–6
religious diversity
      and exclusivism, 216
      and inclusivism, 219, 221
      and pluralism, 216, 218
Riemann hypothesis, 42, 189
Russell, Bertrand, 35, 38, 41, 130,
      135, 188, 189, 200,
      203, 216

Schrodinger, Erwin, 52, 59, 129, 147,
      200
self-organization
      and theology, 210
      in biology, 28
      in complex systems, 27, 29, 30
Sierpinski triangle, 207
special relativity
      and Einstein's postulates, 67
      and Newtonian mechanics, 67, 68
      and the Lorenz transformation,
            67, 68
      and time dilation, 68

spirituality
      and evolution, 215
      and mysticism, 12
      stages of, 11
string theory
      and black hole entropy, 89
      and nine spatial dimensions, 87
      and supersymmetry, 86
      and the cosmic bounce, 87

theology
      and empirical evidence, 198
      and religious authority, 202
      and scientific knowledge, 196, 203
      as a method of inquiry, 194
      the explanatory power of, 199
Trinity, 166, 171
truth
      and propositional logic, 172
      and quantum logic, 173
      and religious teachings, 216
      intrinsically unknowable, 151
      practically unknowable, 146

uncertainty
      decision making under, 11
      Heisenberg's principle, 54, 55
      in physical processes, 52, 147
universe
      and inflation theory, 73, 162
      and the visible horizon, 73, 150
      cosmological models of, 71
unprovable propositions
      in logic, 37
      in mathematics, 34
      in theology, 156, 157

wave function
      and the Schrödinger equation, 52
      collapse of, 53
      interpretation of, 59
Weltbild, 6, 7, 10

CPSIA information can be obtained at www.ICGtesting.com
Printed in the USA
LVOW051107100812

293717LV00008B/24/P